Thomas Meehan

The Native Flowers and Ferns of the United States

In their Botanical, Horticultural and Popular Aspects: Volume I.

Thomas Meehan

The Native Flowers and Ferns of the United States
In their Botanical, Horticultural and Popular Aspects: Volume I.

ISBN/EAN: 9783337112769

Printed in Europe, USA, Canada, Australia, Japan

Cover: Foto ©berggeist007 / pixelio.de

More available books at **www.hansebooks.com**

THE

NATIVE FLOWERS AND FERNS

OF THE UNITED STATES

IN THEIR BOTANICAL, HORTICULTURAL, AND
POPULAR ASPECTS.

BY

THOMAS MEEHAN,

PROFESSOR OF VEGETABLE PHYSIOLOGY TO THE PENNSYLVANIA STATE BOARD
OF AGRICULTURE, EDITOR OF THE GARDENERS'
MONTHLY, ETC., ETC.

VOLUME I.

ILLUSTRATED BY CHROMOLITHOGRAPHS.

BOSTON:
L. PRANG AND COMPANY.
1878.

PREFACE.

THE want of a systematic, illustrated work on the Flora of the United States has long been felt. Some time ago the author of the present volumes seriously entertained a project for such an undertaking, and even went so far as to issue a prospectus. But the difficulties in the way of the enterprise seemed so formidable that it was thought prudent to abandon it. The difficulties alluded to can readily be perceived. A glance at the vast extent of our country, with its widely differing conditions of soil, climate, and position, is sufficient to convince even the most superficial observer that the task of describing and illustrating its Flora is one which might well cause even the most courageous of botanists to hold aloof. To complete such a work in the lifetime of one man would be impossible, and this consideration was one of the main reasons which determined the author to abandon his project. In this determination he was strengthened by another consideration, which, although of an entirely different nature, seemed to be quite as potent as the first. A purely scientific and systematic treatise on the Flora of the United States, in the sense in which such a work would be understood by the botanist, must necessarily be limited to a small circle of readers, and even in this small circle there would be but few who would care to subscribe to a work, the end of which they might never live to see. While, therefore, such an undertaking was clearly an impossibility from the author's point of view, it was equally evident that no publisher could be found ready to invest in it.

Under these circumstances, the fact that a work on "THE NATIVE FLOWERS AND FERNS OF THE UNITED STATES" is offered to the public may need a few words of explanation.

The plan of the present work differs totally from that of the one previously contemplated. In treating the subject, no attempt will be made to be scientifically systematic, from the botanist's standpoint. Instead of the Flora of the United States, the work will embrace simply a selection of the flowers and ferns indigenous to our country. It will be an *anthology* in the truest sense of the word, and will not aim at anything further than to cull the most beautiful, interesting, and important from among the vast number of plants which grow in the different parts of our country. Again, in order to secure the wide patronage which is absolutely necessary to sustain an undertaking of this nature, it has been deemed advisable not to devote the text exclusively to scientific descriptions, but while making it accurate in this respect, to seek rather, by a familiar treatment of the subject, to lift our native flowers out of the confined limits of pure science, and thus to make the work serviceable and accessible, not only to the botanist proper, but also to the practical cultivator, and to the great body of intelligent people at large.

It must not be inferred from this, however, that the work is absolutely without system. It will be seen that the selection made for these two volumes covers a wide range of country, and offers a number of representatives of leading genera, chosen with reference to their various habits, and to different geographical centres. These volumes are therefore absolutely complete in themselves, and may be said to give a good general idea of the floral wealth of our country. Those who are satisfied with the knowledge thus obtained may rest here. But it is hoped that the more enthusiastic lovers of flowers will welcome the succeeding volumes, which it is proposed to publish after the conclusion of this series. Each of the following series is also to consist of two volumes, and to form a complete whole by itself.

With such a plan of publication settled upon, and with the assistance of a competent botanical artist assured, the author felt no hesitation in again taking up his favorite project, more especially when Messrs. L. Prang & Co. consented to become the publishers. The work of Mr. Alois Lunzer, who painted from life all the plants treated in these volumes, the writer heartily commends, believing it will favorably compare with the best hitherto attempted in this country, both as regards scientific accuracy and pictorial excellence. To extol the merits of the chromolithographic reproductions executed by the publishers would be simply superfluous, in view of the widespread reputation of the firm, and with the plates in this work before the eyes of the reader.

Much of the success of the enterprise is due to the kindness of botanists all over the country, who have furnished specimens of plants from their various localities with the greatest readiness. The author must, for the present, content himself with this general expression of his gratitude, as the list of names to be mentioned is altogether too long for insertion in this place, and as due credit will be given in the text in each individual case. But justice and gratitude both demand that a special acknowledgment should be made, even here, of the many favors received at the hands of the authorities of the Botanical Garden at Cambridge. The unrivalled facilities of this institution have been extended to the writer, and to all those associated with him in the preparation of this work, with an unfailing courtesy, for which it seems almost impossible to return adequate thanks in words.

It only remains that the author should say a word in regard to his own share in the undertaking. As already stated, the present work is not exclusively botanical in its character, but is intended to be a contribution to general intelligence. American botanists have done their task so well, that there is scarcely room for even an illustrated work with botanical aims alone. Indeed, but for the labors of Professor Gray, Professor Wood, Dr. Chapman, Mr. Sereno Watson, and other botanists still living, and of the many who have gone before, the work could not have been undertaken at all. The author's task, therefore, has been chiefly to point out the lessons which their labors teach. They have sown the seed, — he shows how to gather the crop. He may not have told all that might have been said; but he believes enough has been brought together to lend fresh interest to the even more than twice-told tale of our native flowers.

THOMAS MEEHAN.

GERMANTOWN, PHILADELPHIA, May, 1878.

CONTENTS OF VOLUME I.

	PAGE
TRADESCANTIA VIRGINICA.	
Spiderwort .	1
GEUM TRIFLORUM.	
Three-flowered Avens	5
GELSEMIUM SEMPERVIRENS.	
Carolina Jasmine, or Yellow Jasmine	9
POLYPODIUM INCANUM.	
Hoary Polypody .	13
VIOLA CUCULLATA.	
Common Blue Violet .	17
ANEMONE NEMOROSA.	
Wind-Flower, or Wood Anemone .	21
AQUILEGIA CHRYSANTHA.	
Golden Columbine	25
PACHYSANDRA PROCUMBENS.	
American Thick-Stamen	29
HELONIAS BULLATA.	
Stud-Flower	33
CAREX STRICTA.	
Tussock-Sedge .	37
CUPHEA VISCOSISSIMA.	
Blue Wax-Weed .	41
THALICTRUM DIOICUM.	
Early Meadow-Rue .	45
ANEMONE PATENS, var. NUTTALLIANA.	
Nuttall's Pasque-Flower .	49

CONTENTS OF VOLUME I.

	PAGE
ORCHIS SPECTABILIS.	
Showy Orchis, or Preacher in the Pulpit	53
SYMPLOCARPUS FŒTIDUS.	
Skunk Cabbage	57
PEDICULARIS CANADENSIS.	
Common Wood-Betony	61
ERYTHRONIUM AMERICANUM.	
Yellow Dog-Tooth Violet	65
PHLOX SUBULATA.	
Moss-Pink	69
SAXIFRAGA VIRGINIENSIS.	
Early White Saxifrage	73
ARCTOSTAPHYLOS UVA-URSI.	
Bear-Berry	77
TEPHROSIA VIRGINIANA.	
Virginian Goat's Rue, or Hoary Pea	81
SEDUM NEVII.	
Nevius' Stone-Crop	85
PLATANTHERA FIMBRIATA.	
Great Fringed Orchis	89
LIMNANTHEMUM LACUNOSUM.	
Floating Heart	93
HOUSTONIA CÆRULEA.	
Bluets	97
VIOLA PEDATA.	
Bird's-Foot Violet	101
CALLA PALUSTRIS.	
Bog-Arum	105
EUPHORBIA COROLLATA.	
Flowering Spurge	109
POTENTILLA FRUTICOSA.	
Shrubby Cinque-Foil	113
LINUM PERENNE.	
Perennial Flax	117
XANTHOSOMA SAGITTIFOLIA.	
Arrow-Leaved Spoonflower	121
CASSANDRA CALYCULATA.	
Leather-Leaf, or Cassandra	125

	PAGE
VIOLA SAGITTATA.	
Arrow-Leaved Violet .	129
GERARDIA PEDICULARIA.	
Fern-Leaved False Foxglove	133
CALOCHORTUS LUTEUS.	
Yellow Pretty-Grass .	137
IRIS VERSICOLOR.	
Blue Flag .	141
POGONIA OPHIOGLOSSOIDES.	
Snake-Mouth	145
CLEOME PUNGENS.	
Prickly Cleome, or Spider-Flower	149
ACTINOMERIS SQUARROSA.	
Squarrose Actinomeris	153
CLAYTONIA VIRGINICA.	
Spring-Beauty, or Notch-Petalled Claytonia	157
ASPLENIUM TRICHOMANES.	
English Maiden-Hair, or Dwarf Spleenwort	161
ANEMONE CAROLINIANA.	
Carolina Anemone	165
ROSA CAROLINA.	
Swamp Rose	169
PACHYSTIMA CANBYI.	
Canby's Mountain-Lover	173
SPIRANTHES CERNUA.	
Drooping-Flowered Ladies' Traces .	177
PHLOX REPTANS.	
Crawling Phlox .	181
CHRYSOPSIS MARIANA.	
Maryland Golden Star	185
IRIS VIRGINICA.	
Boston Iris .	189

TRADESCANTIA VIRGINICA.

SPIDERWORT.

NATURAL ORDER, COMMELYNACEÆ.

TRADESCANTIA VIRGINICA, L.—Leaves lance-linear, elongated, tapering from the sheathing base to the point, ciliate, more or less open; umbels terminal, sessile, clustered, many-flowered, usually involucrate by two leaves; plant either smooth or hairy, with flowers of blue, purple, or white. (Gray's *Manual of the Botany of the Northern United States.* See also Wood's *Class-Book of Botany,* and Chapman's *Flora of the Southern United States.*)

THE "Spiderwort" was one of the first of our native flowers to find a home in England, having been carried to that country from Virginia by the younger Tradescant, according to Parkinson, before 1629. Prof. Gray maintains that the genus is dedicated to the elder Tradescant, who was gardener to King Charles I, but other writers say it was intended to commemorate in the name the services of the younger as well. Before Tournefort and Linnæus had made botany simple by reducing the Latin names given to each plant to two, the generic and the specific, or in other words, the noun and its adjective, Latin names of a much greater length had been applied to many plants, and our plant on its introduction to England was accordingly called *Phalangium Ephemerum Virginianum Johannis Tradescanti.* The contrast between the old and the new name will show how much we have gained by the innovation of Linnæus, although there are still some persons who think botanical names hard to learn. It is from the name *Phalangium,* however, that our plant has been called "Spider-wort," and not "because the juice of the plant is viscid and spins into thread," as suggested by Prof. Wood. Pliny speaks of *Phalangium* as a venomous spider, the bite of which was said to be poisonous, and the same name, and also *Phalangites*, was given to an herb which would cure the spider's sting. Those who have made spiders a subject of special study, notably the Rev. Dr. McCook, believe that there is more dramatic poetry than honest prose in poisonous spider stories, and that the majority of spiders are entirely innocuous, while the few which may be venomous are but slightly so. They have, of course, no stings, but articulated jaws, by which, if at all, they misbehave themselves. However, we are but dealing with the past. The ancients believed there were those who were stung, and that their *Phalangites* was the remedy. We are told that "the roots being turn'd up with new ale and drunk for a month together, it expels poison, yea tho' it be universally spread through the whole body." This must, however, have reference to some other plant to which the same name was applied, apparently a sort of lily allied to *Anthericum*, with which, in the then condition of knowledge, the *Tradescantia Virginica* was wrongly associated. But it fully accounts for the English name "Spiderwort," *wort* being the old Saxon name for "plant." Our true *Tradescantias* are not known to possess any medicinal virtues.

The French common name of the plant is *Ephemerine de Virginie*, taken, as we may readily see, from the early Latin name given it by the English authors. In many parts of our country it has received the name of "Starflower," and even (in Minnesota for instance) "Star of Bethlehem"; but as these names are not only inappropriate, but are also applied to so many other flowers, it is best that they should be dropped for "Spiderwort." The French *Ephemerine* is a very good name, for the flowers remain open but a single day, although there are others ready to take their places in long succession. The poetic sentiments associated with flowers are often far-fetched, but as emblematic of "transient happiness" the "Spiderwort" is appropriate. Says Byron,—

"There comes
Forever something between us and what
We deem our happiness,"

and this could well be echoed by this flower. It is hardly called to the enjoyment of the light of day before its doom is sealed, and it becomes

> "Like a frail shadow seen in maze,
> Or some bright star shot o'er the ocean."

The flowering of the plant is of great interest to the close observer. The buds in the umbel are recurved; just before they flower they become erect, and after fading they bend down and perfect their seed, although sometimes, as Dr. Darlington remarks, failing in that particular. Under a lens the stamens exhibit remarkable beauty, being clothed in the lower part with long, jointed hairs, looking like threads of the richest twisted silk. The kidney-shaped anther, with its golden tint, hung to the filament by the slenderest of connectives, will also attract attention.

The flowers of our plants are found of many beautiful colors. The most common color is a reddish violet, but a pale rose, as well as a deeper rose, and vermilion, carmine, light purple, and white are by no means uncommon in gardens. Those we picture are from specimens gathered for us by Mr. Sternberg, near Fort Hays, in Kansas. The smooth forms of the stalk and leaf will occasionally come out from the same stock with the hairy forms; the smooth forms, however, usually prevail in the East. In gardens the flowers are often found double. There are few plants more deserving the florist's attention. It is a remarkably easy plant to cultivate. When once in good garden soil, it will take care of itself, and continue year after year, in spite of accidents, which, as every cultivator knows, seem successively to destroy more delicate species; yet it does not spread annoyingly, as some do, but waits for the gardener to divide the root stocks when he desires more plants. The tendency to vary, already noted, and to produce double flowers, shows how easily improvements might be directed by a skilful hand. Even as we find it, it is one of the best border plants we have. It is in flower most of the month of May.

The western boundary for the "Spiderwort" seems to be formed by the Rocky Mountains. The writer has found it on the foothills near Pike's Peak, and it is reported to have been met with in higher elevations. Its chief home appears to be from Florida northwestward, not favoring much the New England States. It varies very much in its choice of location. In the East, we usually find it in low meadows, — even some that are quite wet, — among the grass. On the prairies, it is found in much drier places as a rule; while, in Colorado, I have found it in dry sand, and one would almost class it there with plants which love aridity. It is rare that we find such a happy disposition among vegetable beings.

EXPLANATION OF THE PLATE. — 1. The most common form found in Kansas.—2. Smooth form (leaf from a flower stem). — 3. Varieties of color. — 4. Base of the plant, with fibrous roots.

GEUM TRIFLORUM.

THREE-FLOWERED AVENS.

NATURAL ORDER, ROSACEÆ.

GEUM TRIFLORUM, Pursh. — Villous; stem erect, about three-flowered; leaves mostly radical, interruptedly pinnate, of numerous cuneate, incisely dentate, subequal leaflets; bractlets linear, longer than the sepals; styles plumose, very long in fruit; stems scarcely a foot high, with a pair of opposite laciniate leaves near the middle, and several bracts at the base of the long, slender peduncles. (Wood's *Class-Book of Botany.* See also Gray's *Manual*; Torrey and Gray's *Flora of the United States;* Watson's *Botany of Clarence King's Expedition;* and the *Botany of the California Geological Survey.*)

BY old English botanists the plants we now know as *Geum* were called "Avens." An old author, writing before the time of Linnæus, says, "The Avens, for all that we can learn, was unknown to the Greeks, and therefore we can furnish you with no Greek name for it, but it is called in Latin *Caryophyllata*, from the roots smelling like cloves. It is, however, supposed to be the *Geum* referred to by Pliny, the Roman writer, in his History." The name *Geum*, however, is from the Greek *geuo*, and signifies "a good taste," referring to the taste of the roots, as alluded to by the writer aforesaid. All the members of this family have more or less of this aromatic character, and some of the species are used as tea where mild tonics are useful.

Our pretty species is found only in the extreme northeast of the Atlantic United States, but takes a more southerly range as it goes westward. It is found in Colorado, in the Rocky Mountains, in the mountains of Utah, in the Sierra Nevada, and most of the high regions of the Pacific Coast. It was first found by the American botanist Pursh, who named it *G. triflorum*, from its almost always having but three flowers on a stem, as shown in our drawing. But Willdenow divided the genus, and made part into *Sieversia*, in honor of M. Sievers, a Russian botanist, and into this division our plant was placed. The *Geums* of Willdenow had hooked and naked permanent styles to the seeds, while the *Sieversias* have feathery, down-like styles, similar to *Clematis*. Modern botanists, however, rank them all as *Geum*.

As will be seen by our plate, the subject of this chapter belongs to the *Sieversia* section. In old works it is known as *S. triflora*. Its feathery awns afford an interesting study. In the other section of *Geum*, the style is pointed, and, when the ovules are fertilized, falls away. In this section, the styles have not this articulation; and thus, after fertilization, they continue to grow, and eventually become the pretty, feathery heads we find them. The laws which regulate these differences are still obscure, and the subject offers an inviting field of study to those who love to pry into the mysteries of plant-life. We may note that, in a general way, the law which decides these peculiarities generally influences, in some manner more or less similar, all related parts. For instance, in the section of *Geum* with pointed styles, we find, as the seeds or carpels grow, the remains become curved, and give a hooked character to the seeds; and in these cases the sepals or calyx leaves are inclined to recurve or become reflexed also. In our plant, the awns grow erect, there is no recurving tendency, and the sepals and petals follow the same course. This is, perhaps, to be expected as the result of morphological law. If, as we must believe, the calyx, corolla, stamens, and pistils are but the leaves of the plant successively changed into these organs, the unity of the law, as affecting behavior, may be at once suspected. Even when appearances are against this proposition, we may often find that, though seemingly divergent, they are essentially the same.

The chief beauty of the "Three-flowered Avens" is in the rosy red parts of the inflorescence, which gradually deepens up from the green-feathered foliage. The long, slender, involucral bracts and the colored sepals constitute all we can popularly call a flower. The pale pink-white petals show just beyond the calyx, but, so far as the writer has ever been able to note, do not expand sufficiently to be more visible. Unless examined closely, the petals would be rarely seen. The flowers are at first nodding, but at length become erect.

We cannot but think, however, that true taste will see in the foliage very much to admire. The finely cut leaf is suggestive of the much-admired fern, and indeed, in this respect, it is superior to many of that family, but it wants the delicacy of texture which, as much as elegance of form, gives the fern so much beauty; still its rough and heavy character is in harmony with its position as a flowering plant. It is doubtful if the fern, as a rule, would look as well as it does if it had blossoms like other plants upon it. The leaf of the "Three-flowered Avens" is elegant, but it is the elegance of the cultured gentleman, and not of a "lady fair." To those engaged in ornamental designs the leaf affords a good study. Where the idea of combined strength and delicacy is required it would be very appropriate. Neatly pressed, dried, and arranged in a rosette, the leaves make a pretty ornament in leaf albums.

The "Three-flowered Avens" seems to grow very well in gardens, where it forms a neat little stalky bunch of about six inches high. The flower stems do not extend much beyond the leaves, and the blossoms open about the end of May.

We are indebted to Professor Sargent, of Harvard, for the specimen from which our plate was made.

EXPLANATION OF THE PLATE. — 1. The whole plant. — 2. Mature head, with awns.

GELSEMIUM SEMPERVIRENS.

CAROLINA JASMINE, OR YELLOW JESSAMINE.

NATURAL ORDER, RUBIACEÆ (according to Chapman, in *Flora of Southern States*. LOGANIACEÆ, according to Lindley, De Candolle, Asa Gray, and other authors).

GELSEMIUM SEMPERVIRENS, Aiton. — Flowers dimorphous; calyx five-parted, persistent; corolla funnel-shaped, five-lobed, the lobes rounded, emarginate, spreading, quincuncial in the bud, the sinuses impressed; stamens five, inserted near the base of the corolla; anthers oblong-sagittate, extrorse; styles united, filiform, partly persistent; stigmas four, linear, spreading; capsule oblong, two-celled, compressed, opening septicidally to the middle, and loculicidally at the apex, each valve tipped with the persistent base of the styles; seeds several, oval, flat, winged, obliquely imbricated in two rows; stem twining, woody; leaves opposite, lanceolate or ovate, short petioled, with minute stipules, evergreen. (Partly from Chapman's *Flora of the Southern United States*. See also Gray's *Manual of the Botany of the Northern United States*.)

THIS is a distinctively American plant. The genus consists of only this single species, and it has no very close relations outside of the American continent. Willdenow regarded it as a species of *Bignonia*, or Trumpet-flower. Without going into details, however, the student would at once see it did not belong to the Bignoniaceous order by the four stigmas, as all Bignoniaceous plants have the single style, terminated by two thin stigmatic plates, which are sensitive, closing slowly when touched. The nearest allies to the *Gelsemium* are the *Spigelia*, a very pretty, hardy, herbaceous plant, but of poisonous qualities; and two less known genera, *Polypremum* and *Mitreola*. These form a little tribe of exclusives, all of this continent. Our plant is known in the Southern States as "Yellow Jessamine," according to Gray and Chapman, but all those whom we have heard speak of it call it "Carolina Jasmine." It is a woody twiner, with evergreen, willow-like leaves, rambling over bushes and low shrubs, and often ascending trees of considerable size.

It is one of the earliest of spring flowers. The specimen from which our drawing was made was gathered in Florida, in January, and we have seen the plant in flower everywhere in Mississippi in March. The flowers are deliciously scented, and fill the atmosphere with fragrance for long distances around. It is singular that Catesby, who wrote a history of the Carolinas, should say that the plant was not an evergreen. Owing to this error the name of Michaux, *Gelsemium nitidum*, was adopted by De Candolle, but this name is now generally dropped for the one we have chosen, and we refer to it here only that readers may not suppose there are two species under these names. It may be that sometimes the plant drops its leaves. It is subject to "notions," for Nuttall says he found near Savannah a kind which was utterly scentless, a rare peculiarity in a flower that is usually so sweet. This peculiarity may, however, have some relation to its dimorphic condition, a character first pointed out by Professor Asa Gray, in Silliman's *Journal*, in 1873. By this is meant that some flowers have the pistils longer than the stamens, while others have them shorter. In such cases it often happens that the short-pistilled flowers do not seed, their only use seeming to be to furnish pollen for the more perfectly pistillate individuals, and varying odor may go with these varying states.

Notwithstanding its beauty as a climber, and the sweetness of its golden flowers, the Carolina Jasmine possesses qualities dangerous to the ignorant, though of great value to the intelligent medical practitioner. Dr. Peyre Porcher tells us that during the war between the North and the South, when medicines in popular use were cut off by the blockade, this plant was commonly employed as a narcotic. The expressed juice was found to produce insensibility to pain, and yet without stupor. Overdoses, however, produced unconsciousness and death. Dr. Porcher says that the plant is gradually advancing northwards, and speaks of it as having "reached Norfolk," as if on a travelling excursion. Where its starting-place was does not appear. It is very common in Mississippi, Louisiana, and Florida, and thence up along all the seaboard States to Virginia, and De Candolle says it is found in Mexico. If, however, Mexico was its original home, it hardly reached us by what is now the "overland route," for it does not appear to be found in Texas, nor have we any record of it from any place west of the Mississippi River.

Though called "Jasmine" in the South, it has no botanical relationship to the genus *Jasminum*, or true Jasmines. The fragrance simulates the real Jasmine, and naturally suggested the name to the Spanish settlers. Several writers tell us that the Italians call the Jasmine "Gelsomino," and that neighboring nations corrupted this to "Gelsemine." This gives us the origin of the botanical name *Gelsemium*, and even this is occasionally written *Gelseminum*. *Jasminum* is said to have been derived from the Arabic name of the plant, "Jasmin," which is applied to it with slight variations of form in all European countries. It is possible that it may have a relation to a Greek word of similar character meaning "healthfulness," of which the grateful perfume is eminently suggestive.

The true Jasmine — we write Jasmine in preference to Jessamine, so common with American authors — has long had a place in poetry. It is generally regarded as suggestive of amiability. This refers, no doubt, to the white and not to the yellow kinds, for the latter are of a rather obtrusive color, and have not the fragrance which true amiability throws around freely wherever it breathes. But in the white kinds this sentiment finds a fair expression, for while the rather small flowers are not obtrusive, yet no one can fail to notice them, and recognize their modest worth. It is this particular expression of the European Sweet Jasmine, no doubt, that inspired the lines of Fanny Osgood:—

"'Thy heart is like a Jasmine bell,
It yields its wealth of feeling."

Our Carolina Jasmine, however, speaks not to us in this language. There is

"the perfume from the blossom's cell
On every zephyr stealing";

but the deep, grassy green leaves and rich golden flowers speak rather of a rollicking joyousness that spring has come,—a joyousness that finds no bashfulness in its expression, but is rather anxious that all the world should know the good floral season is close at hand. Mrs. Sara J. Hale is the only one of our American poets, that we know of, who refers especially to our native Jasmine,—if Jasmine it is to be. She seems to have it in view speaking of a character drawn by N. P. Willis, whose native grace and elegance, qualities which are certainly peculiar to the *Gelsemium sempervirens* in its growth and manner of flowering, she describes as follows:—

"The fashion of her gracefulness was not a followed rule,
And her effervescent sprightliness was never learnt at school."

This plant is a remarkably easy one to cultivate, even in those parts of the world where the thermometer falls below zero, if it can only have the protection of a cool green-house in winter. It looks best trained over flat trellises of wire or wood, though it is often grown as a cylinder, or on balloons. For a basket-plant it is admirable, as it blooms in the winter season when room flowers are most highly appreciated, and a single flower is sufficient to scent a whole room. It will hardly do well in the open air farther north than its natural territory, unless with some protection. At the Laurel Hill Cemetery, near Philadelphia, there is a plant growing on a wall among some evergreen ivy, the leaves of which afford it sufficient protection, and through which it pushes its branchlets, with the sweet flowers, in early spring. The cuttings grow very easily, if taken off in early summer while the wood is half ripe, put in boxes of sandy soil, and kept in a partially shaded place.

POLYPODIUM INCANUM.

HOARY POLYPODY.

NATURAL ORDER, FILICES (*Polypodiaceæ*).

POLYPODIUM INCANUM, Swartz. — Fronds leathery, evergreen, veins obscure, sometimes reticulating near the margin, smooth and green above, pinnately parted; the divisions oblong, obtuse, entire. The frond beneath, as well as the stipe, thickly beset with peltate, chaffy scales; fruit dots near the margin. (See Chapman's *Flora of the Southern United States*.)

FERNS which grow on trees are confined to a very few species in the United States, and of all of them the pretty little species, called *Polypodium incanum*, is perhaps the most frequently found here. In Europe, at least in England, the most common fern found growing on trees is an allied species, *P. vulgare*, a native also of the United States, where, however, it is more often found in the clefts and on the ledges of shaded rocks than on trees. It is interesting to note that the species now illustrated, which, as we have just remarked, is allied to *P. vulgare*, seems to have advanced from the south towards the north, while *P. vulgare* has travelled towards the south, so that the advancing colonies have met and intermingled in the northern parts of the Southern States. And here we find that, while in the centre of its range, and where we may suppose was its original home, the *Polypodium incanum* is mostly found on trees, when it meets its northern friend it takes to the same habit of often growing on rocks. As so few ferns in our district grow on trees, as compared with those which we might suppose to be able to adapt themselves to such situations, it is more than probable that, in the order of evolution, ferns growing on trees — epiphytal ferns — are a comparatively late class in the sequence of creative time. Some of the earlier botanists regarded this species as a "parasite," but many modern ones do not regard it as even an epiphyte, in the sense of attaching itself to trees, believing that it grows only among moss or other decaying material which is collected on the trunks and branches. In Mississippi, where I have seen it abundantly on oaks, it was always associated with moss, as is the specimen here illustrated, which was kindly cut for us from the roof of a house in South Carolina, by Dr. Mellichamp. But in Louisiana I have seen it running up the straight trunks of trees, firmly attached to the bark, without a trace of moss. In this situation, the fronds dry and curl up during the hot weather, the whole plant looking brown and dead; but when the spring rains come in April, the curled leaves unfold, and the plant resumes growth where it stopped the preceding year.

In regard to its geographical range, I have been furnished with some very interesting facts by Mr. J. H. Redfield, who has followed its history closely. Inhabiting all parts of tropical America and the islands of the Carribean Sea, the *Polypodium incanum* enters the United States by the way of Mexico, through Texas. The Cumberland Mountains appear to have checked its direct progress, for it seems to have no desire to get up in the cold; it therefore branched off, part of the little army marching round to the west and north, going up the line of the Mississippi, and then taking the course of the Ohio, so as to just reach that state; the other winding round to the east, and then going north along the Seaboard States, reaching as far as the Dismal Swamp and the Natural Bridge in its northern march. In this way the great ridges of mountains form an immense barrier between the eastern and western colonies.

No attempts that we know of have been made to cultivate it in the open air further north, and in view of the great distance from its original home, it would not probably succeed; but, fastened to blocks of wood and moss, so that it could be moved to rooms, cellars, or green-houses in winter, after hanging out under trees in the summer, it would make a very interesting object to grow.

The common name of this family of ferns is "Polypody," from its botanical name as given by Linnæus, *Polypodium*, — Greek words signifying "many feet." This is the ancient name of some fern, and was adopted by the earlier botanists as probably belonging to the genus under consideration, to which the name seemed applicable on account of the many "little feet" which the numerous creeping root-stocks of the original species possessed. "*Incanum*" is Latin for hoary, the leaves when dry exhibiting the scaly under-surfaces, of a dull, silvery color. The "Hoary Polypody" would be a good popular name for this fern.

Our plate shows the plant as usually seen, though the fronds are described by Chapman as being sometimes six inches long. We give an enlarged portion of a pinnule, so that the great beauty of the peltate scales may be readily seen. These little scales give a peculiar interest to this species. The fronds themselves are not, to our taste, as pretty as those of its northern sister, *Polypodium vulgare*, which, with its abundant masses of yellow sporangia, all in very regular order, seems to require no further ornamentation. This species does not appear to have the power of producing spores as freely; and, if we may speak of plants as the poets speak, suppose it was annoyed at the superior beauty of the other, and the effort to rival it resulted in these pretty scales! Darwin, in a like fanciful vein, pictures the ice-plant, —

> "With pellucid studs the *ice flower* gems
> His rimy foliage and his candied stems," —

speaking as if he were describing a not very young gentleman who depended somewhat on jewelry to atone for departed charms. He would, doubtless, regard this effort by our little fern as a similar attempt by one of the other sex! In ferns there are two distinct modes of growth. In the one, the fronds push up in a direct way; in the other, they are circinate, or unfold as in the unrolling of a coil. This latter mode is a very beautiful form of growth, and the artist has happily caught our specimen in the unfolding act.

EXPLANATION OF THE PLATE.— 1. A section of the scaly rhizome, with fronds showing the upper and lower surfaces. — 2. Section of pinnule, enlarged three times, and showing the sori and scales.

VIOLA CUCULLATA.

COMMON BLUE VIOLET.

NATURAL ORDER, VIOLACEÆ.

VIOLA CUCULLATA, Aiton.—Rootstocks thickly dentate with fleshy teeth, branching, and forming compact masses; leaves all long-petioled and upright, heart-shaped, with a broad sinus, varying to kidney-shaped and dilated-triangular, smooth, or more or less pubescent, the sides at the base rolled inward when young, obtusely serrate, lateral, and often the lower petals bearded; spur short and thick; stigma slightly beaked or beakless. (Gray's *Manual of Botany of the Northern States*. See also Wood's *Class-Book* and Chapman's *Flora of the Southern States*.)

FEW flowers are better known than the Violet. Our attention is attracted by it from infancy to old age. As the chosen emblem of Napoleonism it has served many a sadly practical purpose, and it has been the theme of the poet from the earliest times. As *Viola* it was known to the ancient Romans, and the great Linnæus adopted the name as the language of science. In those early times, when poetry and nature were blended so closely together, the Violet was received as especially the emblem of constancy. The

"Violet is for faithfulness,"

Shakespeare tells us, and it was no doubt the popular association — this particular "language" of the flower — that led to its appropriation by the Buonapartes. Its other chief associations have been with sweet simplicity and modest, retiring humility; but these characters dwell chiefly in the European species. When Duke Orsino, in "Twelfth Night," declares that the music he listened to

"Came o'er my ear like the sweet south,
That breathes upon a bank of violets,
Stealing, and giving odor,"

or Perdita, in "Winter's Tale," tells Florizel that "before the swallow dares" come, there are

> "Violets dim,
> But sweeter than the lids of Juno's eyes,
> Or Cytherea's breath,"

they are referring to experiences which no American species of Violet will afford. Our Violets will

> "take
> The winds of March with beauty,"

but except to a small degree, in some species, fragrance is wanting.

The species we now illustrate is the commonest of those found in America, so frequently met with as to bear the distinctive name of "Common Blue Violet." It has been found wild from Arctic America to the Gulf of Mexico, westward in the Rocky Mountains, and across the Sierra Nevada, almost to the Pacific coast. It grows in deep, shady woods, as well as in the most exposed places, but generally where the soil is a little damp. It varies very much, and in consequence the older botanists made many species, with distinctive names, out of what are now regarded as but forms of one. As a general rule, the flowers are of a deeper blue in rich, cultivated soil, or in high places, than in low or swampy ground, in which latter they are often of a lilac tint, and with the petals particularly thin and lank. Our own Bryant undoubtedly alludes to this form when he sings so slightingly of "violets lean," which

> "Nod o'er the ground-bird's hidden nest."

The general characteristic of our flower is that of retiring, contented luxury.

In some specimens the leaves are lobed, while in others they are palmately divided, but these variations in leaves are now known to be so common in vegetation that only secondary importance is attached to them in determining species. The general appearance of a plant — all the characters combined — decides the question.

The large, showy, but scentless flowers of this species appear with the first approach of spring, and often in the fall if that season be mild. Some Violets, and those of this species more especially, have the power of perfecting seeds without making flowers, in the popular sense of the word. Early in the spring we have the complete flower, formed of calyx, corolla, stamens, and pistil; but as the season advances the petals are not produced and the calyx remains closed. The anthers, however, perfect a small quantity of pollen, sufficient to fertilize the ovaries, and seed is produced in this way in abundance. This process, in the Violet under consideration, often goes on when the flower-bud is completely under ground. Many plants are now known to have flowers of this character, and on account of these "secret marriages," as the poets say, are called *cleistogamous* plants. It is interesting to note the transition from one of these conditions to the other in the fall of the year. The cleistogene flowers are most abundant in summer, and are often all that are produced at that season; but towards the autumn, a flower will be found with but one petal, another with two or more, till late in winter, or towards spring, the complete flowers appear. It has already been noted that the pollen in the cleistogene flowers is very small in quantity. A very interesting physiological fact has recently been made public by Dr. Kunzé, of New York. The seeds from these flowers are borne in great abundance, while there are only about twelve pollen-grains in each anther. From this it would seem that a single pollen-grain is capable of fertilizing more than one ovule, — certainly a very remarkable fact, if it should be proved beyond doubt.

Though the Common Blue Violet is so well known, and is naturally so variable, it has not given much to the florist so far; some white and violet-striped ones are under cultivation, but this is all.

There are, however, many marked varieties wild, of which we give a few specimens in our plate, and there is no doubt but that, if attention were turned to watching for variations, and then sowing from those selected, some interesting forms might be obtained.

The spur of the Violet is worth special investigation by the student. Inside the spur (see illustration on the plate) there is a fleshy, lever-like projection, and it would be a matter of interest to know not only the uses of this projection, but also whether the spur is formed to accommodate it. In this case the spur and staminate projection are proportionate; but some Violets have long spurs and short projections. On the other hand, there are Violets without these projections from the stamens, and then there is no petaloid spur. Some have contended that the projection is used as a lever, which, on being raised by an insect in search of nectar, causes pollen to be thrown on the insect's back, and the pollen is then taken to another flower, thus "cross-fertilizing" it; but as in this Violet the spur membrane is so closely fitted to the "lever" that it cannot work, it shows how wholly imaginary these speculations are.

Color is supposed to be a provision of nature to attract insects to flowers for this very purpose of "cross-fertilization." But the student will not fail to notice that bees at least very rarely visit this Violet, though in color it is perhaps one of the showiest of all the subjects of the floral kingdom. Rich ground, if partially shaded, grows the plant to great perfection, and we may often see large tracts of such land, in old, abandoned gardens, as near a perfect

"Sea of blue"

as it is possible to expect from any flower.

EXPLANATION OF THE PLATE. — 1. The plant, showing its short, thick, and somewhat fleshy rootstock. — 2. A flower divested of petals, showing the heel-like projection which proceeds from two of the stamens and fills the spur of the corolla. — 3. Varieties of color occasionally found.

ANEMONE NEMOROSA.

WIND-FLOWER, OR WOOD-ANEMONE.

NATURAL ORDER, RANUNCULACEÆ.

ANEMONE NEMOROSA, Linnæus. — Low, smoothish; stem perfectly simple, from a filiform root-stock, slender, leafless, except the involucre of three long-petioled, trifoliolate leaves, their leaflets wedge-shaped or oblong, and toothed or cut, or the lateral ones two-parted; a similar radical leaf in sterile plants solitary from the root-stock; peduncle not longer than the involucre; sepals 4 to 7, oval, white, sometimes tinged with purple outside; carpels only 15 to 20, oblong, with a hooked beak. (Gray's *Manual of the Botany of the Northern United States.* See also Chapman's *Flora of the Southern United States;* Wood's *Class-Book;* *Botany of the Geological Survey of California;* etc.)

THE classical pronunciation of the generic name of this plant is An-e-mo'-ne, but the accepted pronunciation is An-em'-on-e. The Latins tell us that Adonis, the beautiful son of the King of Cyprus, and the "minion of Venus," was turned into a sort of poppy called Anemone. Others tell us that Anemone was a nymph beloved by Zephyr, and was therefore banished by the jealous Flora from her court, and changed into a cold spring flower. Boreas, however, wooed her, but, still true to Zephyr, who in this strait abandoned her, she would listen to nothing he had to say. Finding that she slighted his attentions, he maliciously continued them until she was half inclined to listen, when, after she had slightly opened her petals, he blew a cold blast and caused the tender flower to fade away. There is a popular impression in Europe that the species we now introduce opens only when the wind blows, and it therefore bears the popular name of "Wind-Flower," and this associates the flower very well with the ancient story. The name *Anemone*, as applied to the whole genus, was given to it, as we are told by Sir William Hooker, from the Greek name for wind, and because many species seem to delight to grow in places exposed to wind. The present species, however, grows in rather sheltered places, and has thus obtained the name of "Wood-Anemone," as well as "Wind-Flower."

The classical allusions we have referred to have been used to advantage by poets, who take our flower as the representative of one forlorn and forsaken, and occasionally introduce it in connection with the rugged banks which now and then line the "course of true love." Herbert Smith refers to

> "The coy Anemone, that ne'er uncloses
> Her lips until they're blown on by the wind";

and Dr. Darwin, in the fanciful "Botanic Garden," has the same story in mind when he says, —

> "All wan and shivering in the leafless glade,
> The sad Anemone reclined her head;
> Grief on her cheek had paled the roseate hue,
> And her sweet eyelids dropped with pearly dew."

The fancy that the flowers of the *Anemone* have turned pale from a happy pink is well based on the varying tints of the flowers. They are sometimes found of a deep rosy hue. The tendency to vary is very marked. Some of the European species furnish the beautiful garden *Anemones*, and there is little doubt that care in selection and seed-sowing might result in producing as varied colors in the American as in the European flowers. The American flowers would have the additional charm of fragrance, as a bunch of our species has a delicate, but delicious perfume.

If the ancients had known a little more than they did, they might have done poetic justice to the wrongs of sweet Miss Anemone by making her cold remains, after her death by Boreas, work to the injury of the whole race of gods and goddesses, for the juices of the plant are very dangerous when taken internally; they are said to be useful in a certain class of immoderate hemorrhages, but are too dangerous in overdoses to be often employed. Even the root, held awhile in the mouth, is said to induce a flow of cold, watery matter from the nose. Linnæus reports that cattle feeding on it, in the North of Europe, get the dysentery. Chiropodists sometimes use the juice to burn out corns, and it is said to enter into some preparations for curing gout and rheumatism.

The *Anemone nemorosa* grows abundantly wherever it is found at all, and has a very wide range. It extends down into the mountains of North and South Carolina, and is also found along the coast of California. On both sides of the continent it proceeds far towards the Arctics, and is equally at home in Europe and Siberia.

It is one of the earliest flowers to put in a spring appearance, and is always welcomed by the most practical as well as by those who read "sermons in stones, and God in everything." Among these last, the eloquent poet, Percival, says, —

> "Beside a fading bank of snow
> A lovely *Anemóne* blew,
> Unfolding to the sun's bright glow
> Its leaves of heaven's serenest hue.
> "'T is spring,' I cried; 'pale winter 's fled,
> The earliest wreath of flowers is blown;
> The blossoms, withered long and dead,
> Will soon proclaim their tyrant flown!'"

Yes, the winter is a tyrant to the flowers! but to the plant which bears it, a true friend. It gives it rest, and the "snowy bank," which the poet loves to see fade from over it, furnished protection and warmth to the little roots as they slept; but with our plate before us, we should not call what appear to be roots by that name. There are few "roots" to the *Anemone* in the winter season. What we find then are underground stems, from which the little root-fibres push forth in early spring. From the end of these stems a leaf starts from the apex after a few warm days, and the plant prepares to run on and make a few inches of underground growth for next year.

If it has not very well prepared itself, it may not flower, and then it appears as in our Fig. 3, — a long, petioled leaf only, with a five-parted blade, when, like similar failures in human life, it may be thankful for the chance to try again another time; and, though a season has been thrown away, it generally manages to make a good flowering specimen the next year. In this case we have a pair of leaves from between which the flower is produced. The little difference between the form of the growth when it is barren and when it succeeds in producing a flower is not of much importance here, but it will help us to understand appearances in other *Anemones* that have distinct root leaves, independent of the flower stalks.

EXPLANATION OF THE PLATE. — 1. The usual white-flowered form. — 2. A rose-flowered variety. — 3. An abortive flower stalk. — 4. Full-face view of an expanded flower.

AQUILEGIA CHRYSANTHA.

GOLDEN COLUMBINE.

NATURAL ORDER, RANUNCULACEÆ.

AQUILEGIA CHRYSANTHA, Asa Gray. — Allied to A. cærulea; tall, two to four feet; flowers deep yellow; sepals lanceolate oblong; limb of the petals a little longer than broad. — (*Proceedings of the American Academy of Arts and Sciences*, Vol. VIII, p. 621.)

THE Columbines are celebrated plants. This, the Golden Columbine, has been definitely known only for a short time. Nuttall, Thurber, Wright, and Parry met with it in their travels through the Southwest; but it was thought to be a variety of another species, until Dr. Gray described it as above. It is a native of New Mexico, Arizona, and Southern Utah.

The family of Columbines is represented in the eastern United States by a single species only, while in the Southwest and West there are several. It crosses the American continent to Siberia, and thence extends by several species into the northern and mountainous districts of Europe.

The name, *Aquilegia*, given to this genus, has not been satisfactorily accounted for. Gray, Darlington, and other botanists say it is from the Latin *aquila*, an eagle, from a fancied resemblance in the long spur-like nectaries to the talons of an eagle; but it is quite as likely to be from *aqua*, water, and *lego*, to collect, in allusion to their pitcher-like appearance. These spurs, however, being generally horizontal, or even erect in some cases, would really be unable to collect much rain; but names are often given as much from fancy as from fact. The spurs are called nectaries, because they generally contain a small quantity of sweet liquid. The common name, "Columbine," is derived from the Latin, signifying a dove; but it takes a great deal of imagination to see any resemblance to a dove in our species, in which the horns turn outwards. In many of the European and Asiatic forms, however, the horns are short and bend inwards, and there is a sudden thickening at the end of the horn. The ancient artists, as Dr. Prior tells us in his "Popular Names of British Plants," loved to picture doves feeding together in peace around a dish, and if we set one of the dove-colored Old World forms on the ground, with the horns uppermost, it has exactly the appearance of one of these old-time dove dinner-parties. Darwin, in his notes to the "Botanic Garden," a fanciful old work published seventy years ago, in which plants are endowed with the attributes of animal life, tells us the resemblance is to a nest of young doves, fluttering and elevating their necks as the parent approaches with food for them; but as the dove has but two young at a time, the nest full would be rather slim, and Dr. Prior's explanation is more probable.

Though there is nothing of the dove in the shape of our species, those who love to trace resemblances to animate nature in these inanimate things will see in it a fair likeness to some other bird, indeed a much closer resemblance than can be traced in the *Espiritu Santo*, the "dove plant" of the people of Panama. Take, for instance, the central petal on the left-hand flower on our plate. The anthers might represent a spreading, feathery tail; the petal, the back; the two sepals, a pair of wings; and the long nectary, terminating in a point, the neck and small head.

Some of the poets have dedicated the Columbine to folly; but there is nothing known, either in legend or in history, which couples the name with it, nor is there anything suggestive of such a sentiment in the plant itself. In some passages of an old play by Chapman, written about the year 1600, called "All Fools," and referred to by Mr. Ellacombe in the "Garden," there occurs this passage: —

> "What's that — a Columbine?
> No! that thankless flower grows not in my garden."

But in what particular respect it is supposed to have committed the folly of being thankless does not appear. Another old-time poet, Browne, tasks it with "desertion":—

> "The Columbine in tawny often taken,
> Is then ascribed to such as are forsaken."

In this case it is probable that tawny varieties were seldom seen; and when one did appear, it seemed all alone, deserted, as it were, by its dove-colored friends, and therefore those "who loved to talk in flowers" might find in this exceptional color an eloquent speaker.

Ophelia remarks in "Hamlet,"

> "There's fennel for you, and Columbines,"

and in this might have implied both folly and desertion. It is remarkable that with so extensive an association of this pretty flower with these unpleasant ideas, it has been impossible so far to find any clew to their origin.

The Columbines afford a great deal of interest to those who are fond of studying the laws of plant life. There is a wide range of varying color and form among them, and yet they seem so nearly related that botanists have great difficulty in deciding on the characters which are to define the species. There is, indeed, a suspicion among some of them that they are all merely varieties; that is to say, departures, at no very distant date, from one primordial form. In cultivation Mr. Josiah Hoopes, of West Chester, finds that the European species and those of America readily intermix when growing near each other, the pollen being carried to and fro, either by insect aid or by wind; and some botanists contend that the sweet liquid in the nectaries is secreted by the plant for the especial purpose of inducing insect agency in cross-fertilization. The case with which the varieties or species break up when near each other in this way is the more remarkable from the fact that in their native places of growth each kind is particularly true to a uniform type, variations being rarely met with.

Since the introduction of the Golden Columbine into England it has been taken in hand by the hybridizers, and it is reported that many beautiful varieties have been raised in this way. It is not too much to expect that in time we shall have as many pretty garden varieties of Columbines as there are varieties of Dahlias or Chrysanthemums; for, besides the numerous shades of color which will arise from the mixture of yellow with the various colors already existing in English gardens, we may also look for flowers with an increase in the number of their petals, or, as it is technically called, of different degrees of doubleness, as the anthers very readily turn to petals in our flower.

The Golden Columbine continues in flower longer than any other species we have in cultivation. It is easily raised from seeds and by dividing the roots. The seeds should be sown as soon as ripe, when many of the plants will bloom the next year. If the seeds are not sown till spring the plants never flower till the year following.

PACHYSANDRA PROCUMBENS.

AMERICAN THICK-STAMEN.

NATURAL ORDER, EUPHORBIACEÆ. (*Buxaceæ* of Muller in De Candolle's *Prodromus*.)

PACHYSANDRA PROCUMBENS, Michaux. — Flowers monœcious, apetalous, spiked; calyx bract-like, four parted; sterile flowers numerous; stamens four, with club-shaped exserted filaments; fertile flowers few, at the base of the sterile spike; ovary three-celled, with two ovules in each cell; styles three, thick, recurved; capsule of three one-celled, two-seeded, two-valved carpels. Chapman's *Flora of the Southern States*. See also Gray's *Manual* and Wood's *Class-Book*.)

THE character of the genus, *Pachysandra*, only is given in the above botanical description, as there is only the one species, *P. procumbens*, known in the United States. Indeed, there is but one other species known anywhere, and that is, singularly enough to one who has not studied geographical botany, a native of Japan. It is, however, not uncommon to find isolated species in the Atlantic States of this continent, with corresponding allies in Japan. These are usually of genera represented by a limited number of species, and the phenomenon suggests that there may have been geological disturbances wiping out what probably were the great centres of vegetable families, and leaving only the few scattered outposts on the boundaries. The nearest link now left in the chain of relationship is the common Box of our gardens, although the superficial observer will fail to see much in common between the two. Any one, however, who will compare the flowers of the Box with those of *P. procumbens* must see how nearly the structures correspond. In the Box the spike is very closely drawn together, so as to form a dense head. The lower flowers are all male, with four stamens in each flower, and the female flower, with its three stigmas, terminates the head-like spike. In *Pachysandra* the spike is long drawn out, the male flowers occupying the upper portion, while the female flowers (generally two), with their three stigmas, are at the base. In our plant the stamens have remarkably thick filaments, and this suggested its botanical name, *Pachysandra*, which is the Greek for "thick stamen." The plant seems to have no common name. It may, perhaps, be admissible, therefore, to adopt the translation, and in contradistinction to the Japan species, to call this the "American Thick-Stamen."

Our plant is recorded by botanical authorities as inhabiting woods in mountain districts from Virginia and Kentucky southward to Western Florida; but we seldom find it referred to by local authorities, and it is rarely met with in collections of dried plants made in the South. It is, perhaps, confined to districts out of the usual line of travel. For a plant with a chiefly southern range, it is a very hardy one, for it has been found to endure the winters unprotected in the gardens of most of our Northeastern States. Though, according to the descriptions of authors, it grows naturally in woods, where it may have shade in summer and the protection of leaves in winter, it nevertheless thrives very well in open garden borders without any covering in the winter season. It is very much prized by the lover of curious flowers, not only for the peculiarity of its structure and the earliness with which it blossoms, but for its delicate fragrance. The frost is scarcely gone before it is in blossom, but so inconspicuous is the whole plant that but for the sweetness of its flowers, which attracts insects to it in immense numbers, it might easily be overlooked. Bees from long distances find out the flowers and do homage to their sweets. Indeed, we know of no flower to which the idea of modest worth is more truly appropriate. Many a "wee little thing" possesses "blushing" beauty which has to be sought for among the grass. This does not blush, — it has no color, — but it is retiring, and yet has intrinsic worth. When American poetry shall have appreciated all the pretty expressions of American flowers, we shall have as much of interest associated with this as with the violets of the Old World. In the mean time we can appropriate for it White's beautiful lines: —

> "No ostentatious wish to seek for praise,
> But still retiring from the public gaze,
> It spreads its sweet beneficence around,
> And by the fame it shuns can but be found."

The flowers are arranged to insure self-fertilization, and this is aided by the visits of insects. As shown in the plate, the male flowers are in the upper portion of the spike, the two lowest being the female ones. The anthers burst a few days after the stigmas are in a receptive condition and the pollen can easily fall on them. The insects in their visits only enter the male flowers, and though they get covered with pollen, never come in contact with the pistils; but the stamens have an articulation by which they are readily detached, and after they have been visited by the insects, they fall and carry the pollen to the stigmas below. The blossoming is generally over by the first of May.

The way in which the plant grows on from year to year offers a very pleasant subject for study, and our artist has caught a pretty phase in the plant's life. It is, to a certain extent, a shrub; at least, it makes a shoot one year from which the flower is to come the next. The leaves remain on the little branch till spring, and until after the flower has matured. They commence to turn color as the young flowers form; at the same time, the plant pushes out its new growth for the next year's work. Thus we have the old leaves with their varied colors, the maturing flowers, and the young growth in regular order. The same succession goes on from year to year, all the older growth dying, and in this manner the plant advances, so that in the course of many years it travels a long distance from the original spot, although at the rate of but an inch or so in a twelvemonth. It makes two buds a year, however, and by that means an immense increase occurs in the course of time.

The scales on the young stem-growth make a pretty feature. It is seldom that we see so many in so short a space. The student, of course, knows that they are but leaves modified. The plant needs no leaves underground, but Nature, in her abundant provision, prepares innumerable elementary parts beyond what ever come to perfection, so that she is always ready to act when the time comes. Sometimes these unformed leaves perform the office of bud-scales, and may protect the flower, but the number is so great that they can never be all needed. The transition from the scale or imperfect condition to the perfect leaf, as we see by the plate, is not gradual, but by one great leap, and this, also, is very common in morphology. The change from one form of structure to another, though each be composed of essentially the same elements, is seldom by gradual approaches.

The seeds of the American Thick-Stamen do not mature till autumn. The plant is, however, never raised from seed in gardens, but is propagated by dividing the root-stocks. It is not found nearly as often in gardens as from its many points of interest it deserves to be.

HELONIAS BULLATA.

STUD-FLOWER.

NATURAL ORDER, MELANTHACEÆ.

HELONIAS BULLATA, Linnæus.—Scape ten to eighteen inches high, rather thick and fleshy, hollow, nearly naked; leaves lance-spatulate, about as long as the scape, one to one and a half inches wide; racemes short; pedicels as long as the flowers, colored; flowers purple, segments obtuse, with blue anthers. (Wood's *Class-Book of Botany*. See also Gray's *Manual of Botany of the Northern States*.)

IT is remarkable that while some plants seem to make their way easily, and are found over thousands of square miles of territory, others seem either incapable of wide dispersion, or, if they ever were capable of such dispersion, have lost ground, and are at present confined to very narrow limits. The pretty "wild flower" to which this chapter is devoted is a good example of the plants last mentioned. It is not uncommon in some parts of New Jersey, but beyond these and a few localities in Pennsylvania and Virginia (according to *Gray's Manual*) it is unknown.

To the student this plant is especially interesting as one serving to illustrate a leading division of the great family of lilies, — the *Melanthaceæ*. The flowers belonging to the section given to lilies proper — *Liliaceæ* — have but a single consolidated pistil, though there is normally a three-celled ovary, and the anthers are turned inwards; but those in this section — *Melanthaceæ* — have their styles distinct, and the anthers are turned outwards. There are, of course, other distinctions, but these are the leading ones. Of the genus *Helonias* there are few species, and even these have been placed in other genera by various botanists.

Helonias bullata has a good deal of interest, even to the common observer. The roots are said to be "tuberous" by the describers, but so far as our experience goes there is only a simple fleshy root stock, extending deep down into the ground, from which numerous fibres grow. The plant flowers in May, and the leaves of the old year sometimes remain on during the winter, and then do justice to the description, "about as long as the scape." In our specimen, kindly furnished by Mr. I. C. Martindale from a locality not far from Camden, N. J., the old leaves are gone, and the new ones not fully developed. The "nearly naked scape" is seen to have four or five very small scale-like bracts, and a great peculiarity noticeable is this, that there is neither any sign of scales just under the pedicels (which in fact is not uncommon in many plants), nor of what we might call "decurrence" or running down in the pedicels. The stem, it will be seen, is entirely round, and the flowers come out at right angles, and seem as smooth at the connection with the main stem as if they were pins stuck in. This singular appearance is heightened by the color. There is no shading off, as is general in nature. There is an immediate change from the green main stem to the purple of the pedicel. It is an additional point of singularity that when the flower fades the pedicels become green.

The plant has no common name that we know of. A quaint old English writer says that it "comes from America, where it grows only near Philadelphia, and is called 'Star-flower' by the natives." But this is no doubt a mistake, as the "Star-flower" of the "natives" there, as elsewhere in the United States, is the *Hypoxis*. The generic name *Helonias* is said to be derived from the Greek, signifying a swamp, and is given from the fact that the plant grows in swampy places, though it does not affect these situations more than many other plants; and *bullata* is from the Latin "bulla," which is the name of round "nail-heads" or studded ornaments on castle doors and other objects. We do not know that this flower has ever been taken as a copy for a "stud" or similar ornament, but few could offer a better model. The mathematical proportions of the parts and the harmony of each with the other is very pleasing. The three-lobed and in itself rather heavy ovary is yet in admirable contrast with the light-lined pistils which curve back on the apex of each division of the three cells. The lightness of the petals in comparison with the heavier ovarium is balanced by their double size, and their numerous repetitions of curved lines are relieved by the straight lines of the stamens which stand out above the petals. Then again we see that a pair of petals will make a perfect triangle, either with one whole cell-division of the triangular ovary, or equally as well with the bay formed by parts of two cells or with two whole cells. We have a triangular ovary, three pairs of oval petals forming three more triangles, and the whole forming a regular circular flower. In our drawing the anthers have burst, and are discharging pollen; but before they reach this condition they are of a pale blue, and in this state the flowers would perhaps commend themselves still more in art designs.

In the absence of any recognized popular name, it will not perhaps be inappropriate if we suggest Stud-Flower for it.

We do not know that the plant has been of much use to mankind. Lindley says that a decoction is used in obstructions of the bowels; but it is well to remember that the whole tribe of *Melanthaceæ* is a very poisonous one, and medical experiments with them in unskilled hands will be very dangerous.

We know of no successful attempts at cultivating it. In all the instances that have come to our knowledge the plants dwindled from year to year, soon disappearing altogether. It is quite possible that it is a real swamp-loving plant, and may not find water enough in ordinary garden soil. Many plants have seeds which only germinate where the ground is wet, and they must, of course, unless removed by art, live and die where the seeds sprout; but such plants generally do better when transplanted to drier ground. If this plant is indeed absolutely restricted to swampy ground, it is an exception to rule, and this fact would give special appropriateness to its botanical name of "swamp-lover."

EXPLANATION OF THE PLATE.— 1. Crown of the root with growing spring leaves.— 2. Scape, showing the hollow stem. — 3. Showing the peculiar insertion of the flower. — 4. Full face view, showing the harmonious proportion of lines.

1. 2.

CAREX STRICTA.

TUSSOCK-SEDGE.

NATURAL ORDER, CYPERACEÆ.

CAREX STRICTA, Lamarck.— Pistillate spikelets 2 to 4, cylindric, slender, the upper ones sessile, often staminate at the summit; perigynia ovate, acute, about as long as the lanceolate scale; culms, one to two feet high, rather slender, deeply striate, very acute and scabrous on the angles, leafy at the base, remarkably cæspitose; leaves linear, keeled, often longer than the culm, radical ones very numerous; sheaths striate, sometimes filamentous; staminate spikelets, two or three, often solitary, half an inch to near two inches in length; pistillate spikelets three quarters to one and a half inches long, the lowest on a very short pedicel; scales reddish brown, with a green keel, variable in length and acuteness. (Darlington's *Flora Cestrica*. See also Gray's *Manual*, Wood's *Class Book*, and Chapman's *Flora of the Southern States*.)

GRASSES have mostly hollow and round stems; the Sedges, which resemble grasses, have usually triangular solid ones, and while the former have generally hermaphrodite flowers (flowers with stamens and pistils in the same individual), the Sedges have the genders either in separate spikes, or in separate flowers on the same spike. The origin of the name *Carex* seems uncertain. It is supposed to be derived from the Greek, and to signify " sharp," from the fact that many of the species have such sharp edges to the leaves and culms as to cut the careless handler. But although of Greek derivation, the name is first found in Virgil as applying to this family of plants, and it was adopted by modern botanists just as it stood.

The Sedge Grasses constitute a family numbering hundreds of species, and some of them are found all over the world. Few of them have any beauty to the casual observer, but many of them " will bear examination" remarkably well. The present species always attracts by the earliness of its flowers and by the large and peculiar tussocks it forms in low, swampy grounds. These tussocks are generally a foot high and fully as wide, and very often are all the vegetation that exists to any great extent in swampy places. They are very much assisted in their formation by frost. As the mud and water expand by freezing, the sedge-tufts are lifted by the expansion, and the finer particles of mud settle under them in the early thaw. The tussocks, therefore, do not grow up from the mud as by a stem, but are lifted gradually, and the plant-collector often experiences the truth of this observation to his cost, by finding that they tilt over under his foot, as he steps from one to the other.

A very interesting fact may be noticed in the tussocks in early spring. On the south side the flowers are perfected often a full week before those on the north side. So little warmth is required to bring them forth that the very small difference in the temperature between the north and the south side of the same plant is enough to make this difference in time.

Another interesting observation can be made on the development of the staminate spikes. The stamens burst from their enclosing scales very early in the morning, and by about nine o'clock have opened their anther cells and committed their abundant yellow pollen to the winds. Nothing but dry membrane remains to represent the anthers for the rest of the day. This process commences from the upper part of the spike downwards, and only a few series mature every day. The next, a fresh series, lower down, take their part in this action, and after several days the whole spike has bloomed.

The precise meaning of the division of sexes — the arrangement of female flowers in one head and male flowers in another — is not yet clear to botanists. In these Sedges the pollen-bearing or staminate flowers are usually mature at a time when the pistils of the female flowers on the same spike are not in a receptive condition, and the fertilization of the flower therefore is more likely to be from the pollen of another flower on the same plant, but on another spike, or even from a flower on a different plant. The meaning was supposed to be that it was an arrangement to avoid close breeding; but Mr. Darwin has shown that for any benefit to result from cross-fertilization the two parent plants must be growing under different conditions, which is not the case with the numerous plants of this one Sedge growing in the same swamp together. The true meaning of separate sexes in flowers, therefore, still awaits discovery by some observing student.

The relative positions of the male and female flowers in the Sedges will also interest the observer. In some cases the spike of male flowers terminates the scape; in others the male flowers occupy the lower place; in others, again, they have various places on the same spike. It will be generally noted that this is associated together with lines of nutrition,— those evidently favored by comparative abundance sustaining the female flowers. And this is indeed a natural consequence, for, as vitality exists so much longer in the female than the male flowers, which generally die when the pollen has matured, it is essential that they should have every advantage in this respect. Our present species has the male spike terminal; but as seen in the more mature portion (Fig. 2), the apex of the female spike is sometimes infertile.

In the spring of the year the swamps along the lines of railroad tracks are often burned over, and it is interesting to note that these tussocks, although exposed on all sides to fire, and left as if they were mere lumps of ashes, push out their green blades as if no fire had been about them. It is an excellent illustration of the determination to succeed under severe trials, which is generally successful in those who struggle with the ills of life. In fact, such people are often better for having struggled, and here we have a still further illustration, for the burned-over plants have the full benefit of the salts contained in the consumed vegetation, and push into growth of a healthy, bluish-green, while those that have not been "tried as by fire," and can only make a nutritious use of last year's foliage by gradual and slow decay, grow with a yellowish tint.

The name *stricta* is given from the stiff, upright leaves of early spring, but these droop over gracefully before fall. For the same reason the plant is in some places called "Upright-leaved Sedge," although its best known common name is "Tussock-Sedge."

As a rule the Sedge Grasses are of little value to the human race; cattle exhibit no great relish for them; but this species, when dried, yields very fair hay for cows, though it is not regarded as so nutritious as the true grasses. Its chief use in nature is in aiding swampy ground to gather the soil that drifts from the high land, and make land that will in time sustain a more nutritious growth.

Shakespeare makes Hotspur, in "Henry IV," refer to the marsh-loving character of the Sedge Grass when he speaks of

"The gentle Severn's sedgy bank."

The "Tussock-Sedge" is a native of most of the States east of the Mississippi, except the extreme South, and is also native to Europe.

EXPLANATION OF THE PLATE.— 1. A portion of a tussock in flower in May.— 2. A scape a month later, with achenes or seeds partially formed.

CUPHEA VISCOSISSIMA.

BLUE WAX-WEED.

NATURAL ORDER, LYTHRACEÆ.

CUPHEA VISCOSISSIMA, Jacquin.—Annual, clammy-pubescent; leaves thin, opposite, ovate-lanceolate, long petioled, rough; flowers nearly sessile, borne between the petioles, solitary; petals violet-purple; stamens 12. (Chapman's *Flora of the Southern States.* See also Gray's *Manual* and Wood's *Class-Book of Botany.*)

THIS plant, known by the common name of "Blue Wax-Weed," is not particularly showy, but is sure to attract the collector by its singular structure. The flower has six petals inserted on the calyx, but four of them are mere narrow threads, leaving to the two upper ones the support of all the reputation for beauty the little flower may possess. We are often tempted to believe that the color of a flower is for the purpose of attracting insects. It may be so, and there are many botanists who accept this as the true explanation of the motive for color. Yet it would seem that our plant would have been better served in this respect if all the six petals were of equal prominence in size and color; and it is more than likely, if flowers be really intended to attract insects, and if, as some botanists further contend, certain special flowers are even specially designed to attract special insects, that form is quite as important as color in this respect, and that the variety of form as well as the origin of color may be due to the same cause.

The calyx and stems are somewhat colored, and so help to make the plant attractive. There is a slight swelling on the upper side of the calyx at the base, which gives it a gibbous appearance, and this suggested the botanical name, *Cuphea*, which is derived from a Greek word signifying "curved." The common crape myrtle of our gardens, which belongs to the same natural order of *Lythraceæ*, may give some general idea of the family relations, while the well-known cigar flower, *Cuphea platycentra*, will afford ready means of comparison.

The plant is a most interesting one to study, as showing how very little differences in structure will lead to great diversity in organization. One would not at first sight suppose there was any very close relationship between *Cuphea viscosissima* and the common garden *Fuchsia*, and when the student turns to his text-book of botanical classification he finds them widely separated. But there is really little difference essentially. The calyx in *Cuphea*, as seen in our present species, is not united with the ovary, and the former is in botanical language inferior; but in the *Fuchsia* the calyx is so completely united with the ovary that we see no trace of it until we are beyond the berry, and we say then the calyx is superior. It is only a more complete union of calyx with ovary, that makes what might be an *Onograceous* plant (*Fuchsia*) a *Lythraceous* one (*Cuphea*). Again, if we compare it with a mock orange (*Philadelphus*) or common garden *Deutzia* (order *Saxifragaceæ*), we shall note that these have several pistils, while in *Cuphea* there is but one. Normally, however, there are more, and our plant is to be regarded of so distinct an order simply from the fact that they have been consolidated into one. These little facts help the student much in the knowledge of the relationships of the great families of plants.

Our species has not been found worthy of being admitted to gardens so far, but in a wild state we note a tendency to variation in the size of the petals, and no doubt careful selection might find some forms capable of floral improvement. It is named *Cuphea viscosissima*—the *very clammy Cuphea*—from the extreme viscidity of its exudations, greater perhaps than in any other species of the genus. The plant is, indeed, quite as clammy as the *Drosera*, which is supposed to make use of its glandular hair to catch insects, and in a certain sense to eat them. Our plant is seldom seen without insects adhering to the sticky stems, and it is not at all unlikely that by the aid of the exudation from the glandular hairs the nitrogenous substance of the insect is absorbed and made use of. We have, however, never been able to note the slightest motion in these glandular hairs, as Mr. Darwin observed in the *Drosera*.

Another singular feature to be noted is that, while in most plants the peduncle or flower-stalk arises from the axil or point just between the stem and the base of the stalk, in this case it comes from between the two opposite petioles. This is a feature common to many other *Cupheas*. The flower is probably formed from the whole central growth of the axis, and then subsequently pushed out of position by the development and growth of a new central axis or stem. Another very interesting matter is the way in which the seeds are attached on one side only of the placentæ, and also the bursting of the capsule, with the thrusting out of the seeds before they are mature. The rupture of the carpel and pushing out of the mass of seeds is done with great rapidity, and is worthy of being closely watched by the observer. The seeds have to ripen after their exposure to the open air, — a phenomenon not often met with in the vegetable world. Our artist has shown this feature very well in the plate (Fig. 2).

The geographical relations of this plant are also quite interesting. The home of the genus is in Mexico and Brazil, and there are about a hundred species known, but only two grow within the limits of the United States, and of these, only this one is found to any extent in our country. It may be considered an emigrant from the tropics, and perhaps is still wandering northward. The earlier botanists gave Pennsylvania as its most northern limit, but Dr. Gray, in the later editions of his Manual, locates it as far north as Connecticut. It is mostly confined, however, to the seaboard States, though as we go south it passes the Mississippi and extends down the continent to Brazil.

It has not made its mark in literature in any special capacity.

One single species of the genus has attained some celebrity in Brazil as a febrifuge, but the whole order has little to offer to us so far, but singularity of structure and a petite style of beauty.

Our species is an annual, and is generally found in old fields, or partially shaded waste places.

EXPLANATION OF THE PLATE. — 1. A branch. — 2. Capsule with immature seeds exposed. — 3. Stem, magnified, with captured insect.

THALICTRUM DIOICUM.

EARLY MEADOW-RUE.

NATURAL ORDER, RANUNCULACEÆ.

THALICTRUM DIOICUM, L.—Smooth and pale or glaucous; 1 to 2 feet high; leaves all with general petioles; leaflets drooping, rounded, and 3- to 7-lobed; flowers purplish and greenish; the yellowish anthers linear, mucronate, drooping on fine capillary filaments. (Gray's *Manual of Botany of the Northern States.* See also Torrey & Gray's *Flora of United States,* Chapman's *Flora of the Southern States,* Wood's *Class-Book of Botany,* etc.)

MODERN botanists have been puzzled to account for the derivation of the name *Thalictrum.* Sir William Hooker supposed it might be from the Greek word *thallo,* signifying "green" or "luxuriant"; but those who have succeeded him tell us it is of "obscure derivation." Pliny refers to a plant known in his time as *Thalictrum,* and it is not unlikely that our present botanical name is identical with this old Roman name (the *c* in the modern appellation being simply a misprint for *c*), although the latter is said to have belonged to a plant with some reputation as an "all-heal," while none of the species have any medical virtues, with the exception of perhaps one, which was used as a plaster in some forms of rheumatism and similar troubles, until superseded by *Arnica.* Many an old name has been adopted by the moderns on a still more slender foundation. The common name is "Meadow-Rue," from a fancied resemblance in the leaves to the common garden herb of this name, with "Meadow" as indicating the places in which it loves to grow. The Meadow-Rue proper, however, is one of the European forms, while our species is a denizen of woods or partially shaded places. It grows in the Atlantic States from Canada to North Carolina, and, according to Torrey and Gray, westward to Oregon. Several very closely allied species grow in the Rocky Mountains.

This—the Early Meadow-Rue—has no brilliant colors to recommend it, but its graceful foliage always attracts the early spring-flower gatherer, by whom it is made to do duty for ferns in the ornamental arrangement of the gathered treasures. It is, however, not without interest to the closer student. The sexes are on separate plants in most of the American species, while the European branches of the family have hermaphrodite flowers. These facts have acquired great interest for the botanist since the publication of Mr. Darwin's works. Where the flowers are diœcious,—that is, having the male flowers on one plant and the female on another,—the latter, of course, can only be fertilized by the pollen from a distinct individual, and this would be regarded by Mr. Darwin as so much in favor of the vigor and powers of endurance of the progeny. It might be instructive to students to examine how far inferior the hermaphrodite forms may be in these respects. At first sight it would seem that the hermaphrodite forms of Europe have succeeded just as well, in the struggle for life, as the diœcious ones of this continent; but this should be made the subject of direct examination, for the faithful student of nature takes nothing for granted until he has the facts in detail before him.

The most showy plants are not always the most interesting. They may have beauty and yet teach little. Plain-looking plants, on the contrary, may be very instructive, and this is the case with the Early Meadow-Rue.

In many plants there are leafy appendages at the base of the leaves, called stipules. In general they appear as if they were small leaves, and in a measure distinct from the main leaf. In the class of plants now described there are appearances at the base of the leaves somewhat similar, but they are formed by the flattened and expanded base of the leaf itself. These are not considered stipules by botanists, but are called "dilated petioles." They, however, serve the same purposes as true stipules, and when structural botany shall have been more closely investigated, they may be found to have a similar origin. In our Early Meadow-Rue this spreading out of the base is beautifully illustrated, extending as it does all around, and giving the stem the appearance of having grown through it. Another interesting lesson is derived from watching the development of the flowers up from the leaves through all their stages, and the comparison of the facts as they appear separately in the male and female stalks. Taking our female illustration (Fig. 1), we see that the slender stem bearing the panicle of flowers is but a continuation of the main stalk. If it had been stronger, the branchlets of the panicle, instead of being flowers, would have been leaves or branchlets. A sudden retardation of growth has made flowers of what would otherwise have been leaves. In the lower branchlet, indeed, we see a small leaflet, the arrestation not having been quick enough to make a flower of it. This affords a good illustration of the morphological law, — that the parts of the inflorescence are only leaves and branches modified. But there is still another lesson taught here. By turning to the male flowers (Fig. 2) we see a much greater number of bracts or small leaves scattered through the panicle, and find the pedicels longer than in the female; and this shows a much slighter effort — a less expenditure of force — to be required in forming male than female flowers. A male flower, as we see clearly here, is an intermediate stage between a perfect leaf and a perfect, or we may say, a female flower. It seems as if there might be as much truth as poetry in the expression of Burns, —

> "Her 'prentice han' she tried on man,
> An' then she made the lasses, O,"

at least in so far as the flowers are concerned, and in the sense of a higher effort of vital power.

The Early Meadow-Rue is hardly showy enough for the flower garden, but those who like elegant foliage might find a place for it in some half-shaded corner. It will not be found at all difficult to grow.

EXPLANATION OF THE PLATE.—1. Stalk with female flowers.—2. Stalk with male flowers.—3. Female flower, showing the separate pistils.—4. Male flower with perfect stamens.

ANEMONE PATENS, VAR. NUTTALLIANA.

NUTTALL'S PASQUE-FLOWER.

NATURAL ORDER, RANUNCULACEÆ.

ANEMONE PATENS, L., var. NUTTALLIANA. — Villous, with long, silky hairs; flower erect, developed before the leaves; leaves ternately divided, the lateral divisions two-parted, the middle one stalked and three-parted, the segments deeply once or twice cleft into narrowly linear and acute lobes; lobes of the involucre, like those of the leaves, at the base all united into a shallow cup; sepals five to seven, purplish or whitish, spreading when in full anthesis. (Gray's *Manual of the Botany of the Northern United States.* See also Wood's *Class-Book.*)

WHAT are called "genera" are as much realities as day and night, but it is as difficult, sometimes, to define the limits of the first as of the second; for, in nature, things glide into each other imperceptibly, as day glides into twilight before night comes.

We experience this difficulty in the case of the flower named above. It is an *Anemone;* and yet, in some respects, it borders so closely on *Clematis* that Pursh, one of our earliest botanists, thought it belonged to this genus, and called it *C. hirsutissima*, while others made it into a distinct genus, and called it *Pulsatilla*, which is the Italian common name of a closely allied species, and means, "Shaken by the wind." In *Clematis* there is little tendency to make petals, — indeed, about four petal-like sepals are all that are generally produced, — and the seeds have long, silky tails to them. The *Pulsatillas* make a verticil of sepals, and have no real petals; and the seeds, as in *Clematis*, have silky tails. Dr. Gray, however, as well as other modern botanists, regards those *Anemones*, the seeds of which have Clematis-like tails (*Pulsatillas*), simply as a section of the genus. Our present species, which belongs to this section, has but a single row of large, pale-blue sepals, and these are as silky as the long-tailed seeds. What is called the involucre is a verticil of half-transformed leaves, the intermediate stage between perfect leaves and the sepals.

Our plant was discovered when the section just alluded to was known as the genus *Pulsatilla*, and was dedicated to the great American botanist, Thomas Nuttall; but it was soon found, on a better acquaintance with it, that it was no more distinct from the European and Asiatic form of *A. patens* than *Pulsatilla* is from *Anemone*, and it was, therefore, called "Anemone patens, *var*. Nuttalliana," to indicate that it is considered simply as a *variety* of the same species.

It seems to thrive remarkably well in gardens, and, although not of a bright color, attracts by the large size of the sepals. The earliness of its flowers is also a valued peculiarity. Our drawing was made in the middle of April from a specimen originally brought from the Rocky Mountains. It is said to flower before the leaves come out; but under culture, it has the leaves tolerably well developed before the flowers mature, as seen in our plate.

The *Anemone patens* commences its career as a "wild flower" on the western shores of Lake Michigan, reaches down into Illinois, and then extends northwest by the Rocky Mountains into British America, and, by connection with the typical species, into Siberia. The common name given to this plant is "Pasque-Flower," from the time of its flowering, it being looked for about Easter, or, as it was called in olden times, about the *Paschal* season.

The poets seem to have united in associating the idea of expectation with *Anemone;* not, however, from anything suggestive in the flower itself, but rather from the circumstances of its mythological history. (See *Anemone nemorosa*, p. 21.) The flower is of too transitory a character to be considered the symbol of "expectation," which should rather hope on to the last. Instead of being enduring and constant, our flower soon drops its petals. Its true character is better expressed in the following lines, the author of which we do not know: —

> "There is a power, a presence, in the woods,
>   A viewless Being, that with life and love
> Informs the reverential solitude.
> The rich air knows it, and the mossy sod.
>   Thou, Thou art there, my God!
>   The silence and the sound
> In the low places breathe alike of Thee;
> The temple twilight of the gloom profound,
> The dew-cup of the frail *Anemone*."

The *Anemone patens* is indeed among the frailest of flowers, but it is not often found in the "reverential solitude" of lonely woods. It seems to prefer more exposed situations, and the writer of this never observed in it any nearer approach to a wood-loving habit than the fact that it grows under the scattered pine-trees of the Rocky Mountains.

Among the closely allied species of Europe and Asia many beautiful colored varieties have been found which commend themselves to the cultivator; but in this country we have noted only the one shade represented in the plate, although Don says there is a cream-colored variety here.

The same author also states that the prairie dogs are very fond of the early flowers. This is a singular taste, and we may well wonder, if the report be correct, what they find enjoyable in them, more especially when we consider the bad reputation which the plant had in times gone by. An old writer speaks of it as follows: "The Herb, Flower, or Root being taken inwardly in Substance, are without doubt deleterious, or deadly: It kills by making the Patient look Laughing all the while, whence it obtained the Name of *Apium Risus* (Laughing Parsley). And yet notwithstanding the Standers-by, or lookers-on, may think that the Patient is really a Laughing, or in a Laughing Humour, there is indeed no such thing. It only by its Poisonous qualities hurts the Senses and Understanding, thereby causing Foolishness; and Convulsing the Nerves, especially of the Mouth, Jaws, and Eyes, draws them this way and that way, and sometimes in a manner all ways, making the sick seem to the by-standers as if he continually Laughed, whereas it is only a Convulsive Motion, wringing or drawing of the Mouth and Jaws awry; and so the poor Patient, dying in this Condition, the lookers-on think he dies Laughing, and so report it, when at the same time there is no such matter, but he goes out of the World under the Sense of violent Convulsions, vehement Pain, and the most extreme Torment imaginable."

The ancients, however, also believed the Pasque Flower to have great power against venomous reptiles, and the old writer above quoted reports on this point as follows: "A cataplasm of the Herb or Root is applyed against the Bitings of Mad-Dogges, Vipers, Rattlesnakes, and other Poisonous Creatures; and to places affected with Gout, Sciatica, &c., with admirable success." In our time, Aconite and other Ranunculaceous plants have deprived the Pasque-flowers of all medicinal reputation, but the story so quaintly told by our old author reminds us of another peculiarity in the life of the prairie dog.

It is well known that this animal burrows deep holes in the ground, the earth drawn out in working the burrow forming a little mound at the outlet. The popular belief is that the owl and the rattlesnake make their home in these underground chambers, and that the three animals live together in peace and harmony. This is a remarkable fact, if true, since most snakes regard the young of birds and other animals as desirable delicacies. The writer has, however, often seen the owl on the mounds of the prairie dogs, and it is possible the rattlesnake part of the story may be as correct as the other; but if this is so, might we not say, with as much reason as the ancients usually had for what they believed, that the prairie dogs use the flowers to protect themselves from the bites of their poisonous fellow-lodgers?

EXPLANATION OF THE PLATE.—1. Full-sized plant.—2. Stamens and pistils after the sepals have fallen.—3. Head, with long-tailed achenia.—4. Single achene, or seed with tailed awn.

ORCHIS SPECTABILIS.

SHOWY ORCHIS, OR PREACHER IN THE PULPIT.

NATURAL ORDER, ORCHIDACEÆ.

ORCHIS SPECTABILIS, Linnæus. — Root of thick, fleshy fibres, producing two oblong-obovate shining leaves, three to five inches long, and a few-flowered, four-angled scape, four to seven inches high; bracts leaf-like, lanceolate; sepals and petals all lightly united to form the vaulted galea or upper lip, pink purple; the ovate undivided lip, white. (Gray's *Manual of Botany*. See also Chapman's *Flora of the Southern States*, and Wood's *Class-Book*.)

THE Orchid family is well known as the most peculiar in the vegetable world. In the temperate regions of Europe, Asia, and America, the plants belonging to it grow in the earth; but in the tropics they generally attach themselves to trees and other objects, deriving most of their nutrition from the atmosphere. The flowers, in many cases, resemble living creatures, frequently vying with them in the beauty of their colors and markings; and singularly dependent, in many cases, on insect agency for the fertilization of their flowers. The purpose of the necessity for fertilization by external agency does not seem clear, though many leading botanists believe it is expressly to avoid self-fertilization, which they regard as injurious, but an Australian species closes its flower with a spring and catches the visiting insect, according to Drummond, thus effectually destroying its chances of cross-fertilization, if that were the object in view.

It is indeed difficult to decide on the purpose of Nature in the structure or behavior of plants, or their several parts, because Nature's purposes are never wholly with a present view. We know by geological and other evidences that the plants of the present age are not as plants were in past periods of the world's history. There is an evident purpose that in the future, plant-races shall not be as they are now, and in pursuit of this plan Nature must of necessity have a destructive as well as a preservative policy, and how a plant behaves may not therefore be necessarily for its own good in the sense in which we understand goodness. Yet there is a tendency to question the plant as to the reasons for the phenomena it exhibits, while the questions should really be addressed to an external power which is looking into the future far beyond.

Not only may we ask, Why are these flowers arranged for cross-fertilization? but, Why are they made to simulate so many forms of the animal world? Some have supposed that the resemblance to insects was to attract insects, but it is difficult to understand how the Orchids accomplish this any better than those flowers which have no peculiar form. If there were any design in the relationship between the flowers and the animate forms they represent, it might have been to frighten the insects away, for we rarely see an insect interfering with another while it is at work. Indeed, this point has been actually suggested by one of the poets, in the following lines: —

> "The orchis race, with varied beauty, charm
> And mock the exploring fly, or bee's aërial form."

These remarks are offered that the student may not hastily decide from form, or the arrangements of structure, that the immediate purposes of Nature are clearly manifested. Very often the plant's behavior has a direct relationship to its individual prosperity, but by no means always. Our species has no striking resemblance to any particular insect, but it attracts all lovers of wild flowers by the very pretty contrast of the delicate rosy-pink upper sepal with the large white labellum or lip. The unusually long spur is a striking characteristic.

The name "Orchis" was already in use by the ancients; but with the progress of botany, the species bearing this name have been placed in various genera, so that the one we illustrate is now the only representative of the genus *Orchis*, as established by Linnæus, which we have in the United States, and even this was transferred by Sprengel to *Habenaria*.

Most of our Orchids, that we should call pretty, seem to prefer growing in open places; but this is one of the few which delight in the shade and shelter of the woods, where it is among the later spring flowers to bloom. In Pennsylvania, it is to be found in the early part of the month of June, and probably a little earlier or later, as it grows northward or southward of this. Dr. Gray gives its range as from "New England to Kentucky, especially northward." Botanists generally do not report it as very abundant in any one place. The writer has seldom been able to gather more than a dozen or two on any one botanical excursion, though it is doubtless more plentiful in some places. Dr. Darlington, in his *Flora Cestrica*, speaks of it as being frequent in the rich woods of Chester County, Pa., and as having the common name there of "Preacher in the Pulpit." It seems, however, to have no popular name in other parts of the United States. Dr. Gray, in the Manual, simply translates its scientific name, "Showy Orchis."

Generally speaking, our native terrestrial Orchids are impatient of culture. They will sometimes do well for a few years, but usually disappear in time. This one has not been tried to any great extent, but would no doubt transplant well to places similar to those in which it is found naturally, and might then perhaps spread, and do well. It could be made to succeed, if the same amount of skill were brought to bear on it which the intelligent cultivator gives to the epiphytal species from the tropics.

It is interesting to note that, while most of the true Orchises of Europe have a tuberous root in addition to the fibres, our species has fleshy fibres only. In the foreign species, there are a pair of these tubers, one of the past and the other of the present season's growth, the one growing out of, and seemingly being supported by the other, and at length appearing to draw wholly out the parent's life. On this Darwin, in his "Botanic Garden," has the following expressive lines: —

> 'With blushes bright as morn fair Orchis charms,
> And lulls her infant in her fondling arms ;
> Soft plays affection round her bosom's throne,
> And guards his life, forgetful of her own."

Our illustration is from a Pennsylvania specimen.

A useful starchy product is obtained from the roots of some of the European species of *Orchis;* but our species is of no known value to man, unless, as some good thinkers will have it, mere beauty is as essential as the more material things of life.

SYMPLOCARPUS FŒTIDUS.

SKUNK-CABBAGE.

NATURAL ORDER, ARACEÆ.

SYMPLOCARPUS FŒTIDUS, Salisbury.—Spathe conch-shaped, acuminate; spadix on a short, peduncle-like scape, oval and densely covered and tessellated with flowers; stamens four, opposite the fleshy, cucullate sepals; ovary one-celled; style four-sided, tapering to a minute stigma; fruit an oval, fleshy, berry-like mass coalesced with the base of the persistent sepals and imbedded within the spongy receptacle; seed globular, destitute of albumen; leaves at first orbicular cordate, finally cordate oval, on short petioles; spadix much shorter than the spathe. (Darlington's *Flora Cestrica.* See also Gray's *Manual,* Wood's *Class-Book,* Chapman's *Flora of the Southern States.*)

UNDER the name of "Skunk-Cabbage," the plant we now illustrate is very widely known. It is our earliest flowering plant, and the news of its first appearance is always hailed with delight by those who are anxiously looking for the first flowers of spring. It is singular, indeed, that it appears so early. No matter how deeply the ground may have been frozen in the winter, the first few warm days find the flowers ready to expand. The roots are seldom less than six inches from the surface, and it is quite probable that the pushing buds have grown up in some degree during the winter, thawing their way, as it were, through the frozen ground; for plants are in some respects like animals, and must keep up a certain degree of heat, no matter how low the temperature may be about them. The degree necessary is not, of course, near so high as that required by animals, but it is not probable that the juices of these plants ever thoroughly congeal, and thus the buds are able to keep travelling slowly upwards at comparatively low temperatures. That the parts would die if frozen is shown by some of the earliest flowers. Very often they are in such haste to open that they mistake a few warm February days for the return of spring, and expand only to meet severe weather. In these cases we find the spadix or interior mass of flowers (see Fig. 2) frozen through so solidly that it is with difficulty they can be cut apart, and then they become black and rapidly decompose on thawing. In the spring of 1877, the writer of this noticed plants in full flower in early March that were afterwards subjected for a week to a temperature below freezing point, and part of the time to eighteen degrees below. How little heat is required to bring forth the flower is well illustrated in one of Collinson's letters to Bartram, who sent some plants to England, which Collinson says had "beautiful flowers on them when the package was opened," called out by the mere heat of the ship's hold.

The Skunk-Cabbage can also teach us a good lesson in botanical relationship. Everybody knows the Calla of our green-houses, properly *Richardia Æthiopica*, and many know that it belongs to the *Araceæ* or *Arum* family. The relationship between these two plants will at once be suspected. It is close, but there is some difference. Looking at the Calla, we see the spadix has male flowers along the upper portion, and the female flowers separately below. Our plant has these organs both in the one little flower. They are hermaphrodite, while the true *Arums* are monœcious. The family to which our plant belongs has been separated as *Orontiaceæ* by some, but our distinguished botanist, Dr. Asa Gray, classes it with the *Araceæ*. Indeed, characters founded on sexual organs are unreliable. In the Skunk-Cabbage they are variable. In most of the flowers of the spadix we find four stamens and four sepals, but in the course at the base there are generally five of each, and instances of five stamens with only four sepals are not uncommon in the upper flowers of the head. It is very likely that in some cases the pistils entirely abort, leaving nothing but perfect stamens to represent the flower. We have here a good lesson on the unreliability of these parts in establishing fixed characters in botanical descriptions.

It will also please the student to watch the development of stamens and pistils. If the temperature remains above forty-five degrees for about three days, the stamens will be fully developed in that time, but if only a very little above freezing point, it takes about a week to mature them after the pistil has been fully developed and is ready for pollenization; for the pistil seems to finish its growth before the stamens begin to make theirs. The stigma is a beautiful object under the lens, being capped by a crown of delicate, fringy hairs. The anthers are very large, and soon burst, discharging an immense amount of pollen, not only on their own pistil, but on those below. At the bottom of the shell-like spathe an immense quantity collects, and gives us some idea of the wondrous exuberance of nature.

Again, there is much of interest in this flower in connection with modern theories of the necessity and utility of cross-fertilization. *Araceæ* have dry, dusty pollen, and generally colorless floral envelopes, and they are thought to be cross-fertilized by the aid of the wind. The maturity of the pistil before the stamens in the same flower is also regarded as indicating that the purposes of nature would be better served by the pollen being received by the stigma from another flower. In the case of our species, the spathe coils round the flower-head and protects it from the wind. It might be that the spathe is necessarily coiled to protect the flowers in this dangerous season, and so color is bestowed on it to attract pollen-carrying insects; but there are none of this class at this season. The scent may attract flies, and these do visit the flowers. If the temperature goes suddenly to sixty degrees, as it often does in early spring, even though the thermometer may have been for days below the freezing point, flies will abound. Pollen might possibly be carried by them to the unfertilized pistils, and this would appear so probable that any one delighting in generalizations might take it for granted that cross-fertilization is thus effected; but the student takes nothing for granted when actual observation can be had. The writer of this has never been able to detect the slightest trace of pollen on the stigmas until they receive it from the flowers in their own spathe. Other students may, however, be more successful. This is one of the many unsettled questions that will give a zest to the studies of those who desire to observe critically the development of the flower.

The plant has been called "Skunk-Cabbage" or "Skunk-Weed" from its odor; but this is most marked after being bruised. If one will bend down over a flower and smell before gathering it, there will be little experienced that is disagreeable. The old Swedish settlers around Philadelphia used to call it "Bear Weed." Bears were abundant among them in those days, and it is said that after coming out from their long winter's sleep, they found this early plant a great luxury. It must have been a hot morsel, as the juice is acrid, and is said to possess some narcotic power, while that of the root, when chewed, causes the eyesight to grow dim. Infusions of the plant have been used by some physicians in whooping-cough and dropsy. The plant is found only to the east of the Mississippi, chiefly from North Carolina northwards; and it has no very near relations. Linnæus thought it a *Dracontium*, under which name it is still referred to by comparatively modern authors. Sims refers to it as a *Pothos*, under which designation the student will yet sometimes meet it; but *Symplocarpus* is its now generally accepted name. This is from the Greek, and signifies, "united fruit." If we examine the fruit of the common Indian turnip, we find it a mass of separate (red) berries. In our plant the parts that might have been distinct are so united together as to form but a single, rough, globular mass, in which the seeds are imbedded, and of so peculiar a structure that Nuttall thought the plant viviparous. After separating from the receptacle and becoming scattered through the ground, the seeds are occasionally found by laborers or others when digging in the swampy places where they grow, and are generally regarded by them as petrified corn, and as such have often been brought to the writer.

EXPLANATION OF THE PLATE. — 1. The plant in flower before the leaves are far advanced. — 2. The spathe half cut away to show the spadix. — 3. Longitudinal section of spadix, showing the arrangement of the single flowers on the receptacle. — 4. Individual flowers.

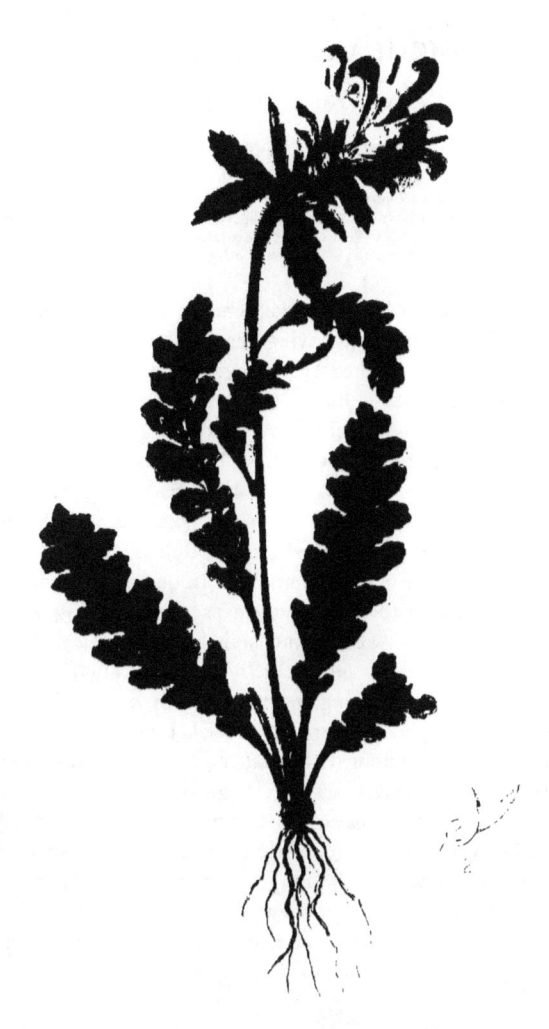

PEDICULARIS CANADENSIS.

COMMON WOOD-BETONY.

NATURAL ORDER, SCROPHULARIACEÆ.

PEDICULARIS CANADENSIS, Linnæus. — Hairy, stems clustered, oblique; leaves lance oblong, pinnatifid; calyx obliquely truncate; upper lip of the corolla with two setaceous teeth at the apex. (Darlington's *Flora Cestrica*. See also Gray's *Manual of Botany of the Northern States*, and Chapman's *Flora of the Southern States*.)

PEDICULARIS is a large genus, over a hundred species which belong to it having been described. Its members are most numerous in the Arctic regions, or at high elevations in mountain districts. Quite a number are found in the Rocky Mountains, and some species grow in the high regions of Mexico. In the Atlantic States we have but two, one of these being *P. Canadensis*, now figured. This has a wide range for a plant whose family relations are so far to the north, as it is found in almost every State, from Canada to the Gulf of Mexico, and extends west to the Rocky Mountains. In our country, however, our species seeks shade from the warm suns by taking to open woods, or getting on rising knolls in swamps or low grounds, where it may have the advantage of a humid atmosphere. It flowers very early in spring, being generally out of bloom and having its fruit ripened before the first of June.

The flowers are amongst the handsomest of our native plants, and the fern-like leaves set off to great advantage the floral beauty. An unusual feature is the great variety in the colors, at least in the specimens generally found in Pennsylvania. The upper portion of the corolla ranges from a light brown to a rich purple, while the lower portions are of a pure white, varying to a light yellow. These natural tendencies to change offer great inducements to the florist to attempt improvements. At any rate, the wild forms can be selected for cultivation. The plants do well when transplanted from their native wilds to our flower borders, if they are not in a situation much exposed to the full sun.

To those who love to watch the various processes of nature in the floral world, the manner in which these flowers are fertilized affords an interesting study. It is difficult to understand from the structure how they self-fertilize, or how they can receive much help from insect agency; and besides, insects will rarely be found visiting them,—at least this is the writer's experience, —and yet every flower seems to perfect seed. There is evidently a field here for further discovery.

The name *Pedicularis* is a Latin adjective, signifying "belonging to a louse." In the northern countries of Europe some of the species abound, one of them, *P. Sceptrum Carolinum*, to such strength and in such abundance that, according to Linnæus, it stopped a horse going at full speed. In these countries the whole family is in bad odor with stock-raisers, from an idea that cattle, and sheep especially, feeding on them become lousy. Like many other old notions in agriculture, this is no doubt a libel on these beautiful flowering plants. But the impression induced Linnæus to give the name to this genus, and from it also comes the English name of Lousewort, *wort* being an old Saxon name for "plant." Americans, however, follow Dr. Gray in calling the plant "Wood-Betony," the "Betony" being from some resemblance to an English wild flower of that name.

The young botanist who attempts to dry plants is generally astonished that, with all his care, this one, admired so much in life, defies all efforts to preserve its colors well. It turns black under the best of care.

Some poets refer to the Betony in connection with "surprising situations or circumstances." This scarcely has reference to our plant; but if it had, the association would be not inappropriate. It is a matter of "surprise" that a flower so beautiful should have received so little poetic attention. Shakespeare, whose genius for observation was so universal, wholly overlooks it.

Perhaps the European species does not strike the observer so favorably as ours strikes us. On the Wissahickon, near Philadelphia, there are rolling banks in the deep shade of woods completely moss-grown, among which the trailing arbutus or *Epigæa* finds a welcome home. In the earliest spring the young go out to seek these beautiful flowers, and they have hardly gathered the last when our *Pedicularis* is ready for the floral harvest.

Perhaps, after all, it is often accident, more than actual worth, which brings some flower popularly forward. As Young says,—

> "But own I must, in this perverted age
> Who most deserve can't always most engage;
> So far is worth from making glory sure,
> It often hinders what it should procure."

We have taken for our picture only a single branch from the root-stock. It is not uncommon to find a dozen or more in an old plant, all in bloom at the same time.

The way in which it pushes up and forms its flower-stems is interesting to the morphological student. When the flower-stem starts to grow, another set of buds begins to prepare for the next year. These buds proceed with their development at the side of those which are now making the flower shoot. The new buds form a tuft of a dozen leaves or so, and remain in that condition till the next spring, when they also throw up a flower shoot. Now this little tuft of a dozen leaves is really the equivalent of a branch. We must imagine a branch with a dozen leaves on it, spread apart so as to have an inch or two of space between each one. Then imagine this branch drawn in, as we draw in the circles of a coil of wire, and we have just the idea of these tufts of leaves. Now when the plant begins to flower, the spiral is drawn out, the leaves are scattered on the stem, and the head is borne upwards; but when the true flowering time is reached, we see that there is a sudden stoppage of this elongating growth, and we have a whole coil of bracts, but little altered from true leaves, forming a verticil under the spike of flowers. We see by this that the leaves had been pretty well developed before the drawing in of the spiral coil commenced, and the lesson taught us by our flower is therefore this, — that the matter of time in the acceleration and retardation of development is the main cause of many of the varied forms found in vegetation. When the accelerated motion precedes leaf development, as in many plants it does, there may be but very small bracts, or even no bracts at all. In most other species of *Pedicularis* the development is regular, and the involucral-like circle of bracteal leaves does not exist.

A further lesson we may gather from the flowers. The bracts — the small leaflets among the flowers — are changed leaves, and the flowers which spring from the axils are analogous to the branches which spring from the axillary bud at the base of the perfect leaf. A flower is, therefore, a modified branch, as the bract is a modified leaf. In many flowers we can trace the relations of the floral parts to leaves and branches; but in this the arrestation has been so severe that we lose all resemblances in the flower, and we cannot tell whether the corolla is made up of a single leaf or several. The attention of the student is directed to this point because here will ultimately be found the full explanation of the reason why flowers are sometimes of very peculiar forms.

Generally, we can tell what form the seed-vessel will assume before the petals or the corolla fade; but in this *Pedicularis*, the capsule continues to grow, and ultimately assumes a sword-like beak, projecting beyond the calyx. (See Fig. 2.) When mature, it opens by a slit on the upper side through which the ripe seeds escape. Altogether it is a very interesting plant to study, as well as a pretty object to look at for those who wish to enjoy only the external beauty of nature.

EXPLANATION OF THE PLATE. — 1. A single branch from a root-stock. — 2. Calyx and mature seed-vessel.

ERYTHRONIUM AMERICANUM.

YELLOW DOG-TOOTH VIOLET.

NATURAL ORDER, LILIACEÆ.

ERYTHRONIUM AMERICANUM, Smith. — Leaves elliptical-lanceolate, pale green, mottled, and commonly dotted with purplish and whitish; perianth light yellow, often spotted near the base; style club-shaped; stigmas united into one. Scape six to nine inches high; flowers one inch or more long. (Asa Gray, *Manual of Botany of the Northern States.* See also Chapman's *Flora of the Southern States,* and Wood's *Class Book.*)

THIS is one of our earliest flowers, being in full bloom in Pennsylvania the end of April and beginning of May, and earlier or later in Southern or Northern States. On this account it received the name of "Yellow Snowdrop" from the earlier settlers in Pennsylvania, who remembered the early-blooming snowdrops of the Old World. Many other common names have been given to it, but "Yellow Snakeleaf" prevailed generally with the last generation, and it commonly receives this name from modern writers on popular botany. The name, however, which seems most in use at the present time, and which, we think, will prevail, is "Yellow Dog-tooth Violet." It varies very much in the markings of the leaves in some localities. Sometimes there are scarcely any spots; then it often receives the name of "Lamb's Tongue." The name "Dog-tooth Violet" is derived from the roots of the single European species, *Erythronium Dens Canis,* which is literally *Dog's-tooth Erythronium.* So great is the resemblance to the canine teeth of the great friend of man, that the roots seem to have had this name among all the old nations of Europe long before it was adopted by science, and indeed long before plants had any botanical names at all.

The resemblance to the violet is rather imaginary; but as the European form, usually white, is often purple in Italy, and blooms about the same time with the violet, the popular name would at least seem to be explicable. The name of the genus is not so well traced. Dr. Gray says, "*Erythronium* is from *Erythros*, Greek for 'red,' which is inappropriate as respects the American species." Prof. Wood seems of the same opinion, as he says that the name is derived from "the color of some of the species." But none of the European varieties have flowers of a color deep enough to suggest such a name. Dr. Darlington believes that the name was from "the purple stains on the leaves." Botanists do not always give the reasons for their names, and we are left to guess at them. The earlier ones delighted in adopting ancient appellations. *Erythronium* occurs in Pliny and Dioscorides, and some of the older botanists thought it had reference to this plant, and so retained it, though the plant referred to by these ancient writers was evidently used in dyeing, which the Dog-tooth Violet could not be. The family has, however, some use in human economy. The powdered root of the European species was once used, with milk, for intestinal worms in children. The root is rather acrid when fresh, but becomes mealy when dry. Rafinesque says fresh roots and leaves, stewed in milk, make a rapidly healing application to scrofulous sores. Dried bulbs, however, lose this virtue. Porcher, the most recent American author on the medical properties of plants, says the bulbs are emetic when powdered, and given in doses of twenty to forty grains.

The Yellow Dog-tooth Violet is found in damp, shaded woods in, we believe, all the Atlantic States, and westward as far as the Mississippi, beyond which it gives place to other species. The order to which it belongs is very small, consisting of perhaps not over half a dozen individuals, even if we include the marked varieties. Its nearest ally in our country is *Lilium*;

the pistil, however, is not three-cleft as in the lilies, but the lobes are united, forming a club-shaped pistil, as shown in our plate, and it also differs in other characteristics. A nearer relation exists between it and the common garden tulip, which has, however, a bell-shaped flower-cup, and a sessile, three-parted stigma. At night our flowers close, opening somewhat as the day advances, but on warm, sunshiny days they recurve as completely as the " Turk's-cap Lilies."

There are, no doubt, many interesting facts in the life-histories of the Yellow Dog-tooth Violet which yet remain to be recorded. In some localities, as already noted, the leaves are not spotted as in other cases. In these instances there seems to be a difference in the disposition to produce seeds, as if the two points went together. Then again in some localities there are immense numbers of small roots with only one leaf, and but a very few — the flowering ones — with two, and it is not known how long it is before a seedling-plant flowers. In the tulip the young roots do not flower for several years, and it may be the same with this.

It bears culture very well, provided it be grown in a partially shaded place; and no doubt, with attention, as many varieties might be raised as have been produced in the tulip.

Though so old a plant in history, the poets seem to have overlooked it, its companion, the violet, having evidently had superior charms for them. But as we have not the tulip with us, and the genus is allied to it botanically, what the poets have said of the one may without much violence be transferred to the other. Holland makes the tulip reflect on its own merits, in contrast with other floral favorites, as follows: —

> " How vain are the struggles for conquest and power
> With golden bud and scented flower,
> Who claim, from their beauty or fragrance alone,
> Their right to ascend the garden throne!
> A graceful form may please the sight,
> And fragrant odor the senses delight;

> Yet if we are judged by our merit, I ween
> The Tulip will soon be the Garden Queen;
> No envy I fear, nor of beauty the frown,
> While the worth of the Tulip can purchase the crown.
>
> "How can the vain Rose ever hope to claim,
> By the verse of the poet, the bright meed of fame?
> Or the pale-featured Lily pretend to enhance
> Her right, as the flower most favor'd of France?
> No favors I boast, though in beauty I shine,
> And variety's garb, ever charming, is mine;
> But my triumph I rest upon merit alone,
> For worth is e'er valued when beauty is flown.
> Then why should I fear either anger or frown,
> While the worth of the Tulip will merit the crown?"

The only incongruity in the application of these lines to our plant is in the line, —

> "And variety's garb, ever charming, is mine."

But, as already remarked, there is little doubt, if zealous improvers would take it in hand, this boasted charm would be our plant's as well. The original tulip of Europe (*Tulipa sylvestris*) is a simple yellow flower, a little larger, but scarcely so showy as this lovely spring flower of our woods.

EXPLANATION OF THE PLATE. — 1. Whole plant with bulb deep in the ground. — 2. Side view of flower, with relative length of pistil and stamens. — 3. Recurved petals as often seen at mid-day. — 4. Capsule soon after the petals have fallen.

PHLOX SUBULATA.

MOSS-PINK.

NATURAL ORDER, POLEMONIACEÆ.

PHLOX SUBULATA, Linnæus. — Stems prostrate, twelve or more inches long, with numerous assurgent branches two to four inches high; leaves subulate, linear, rigid, about half an inch long, cuspidate, crowded, with axillary clusters of smaller ones; corymbs three to six flowered; corolla pink purple, with a dark-purple centre, the tube about half an inch long, a little curved; flowers sometimes white. (Darlington's *Flora Cestrica*. See also Gray's *Manual of Botany* and Chapman's *Flora of the Southern States*.)

THE *Phlox* is an American genus of plants, but was one of the earliest to obtain an introduction to the botanists of Europe. Plukenet, a writer before the time of Linnæus, published a work in London, in 1691, in which he describes it, making it out to be a near relation to the *Lychnis*, for which reason he called it *Lychnidea*. The *Lychnis* belongs to the Pink family, or, as we say, *Caryophyllaceæ*, and there is much outward resemblance of the *Phlox* to it, especially in the seed-vessel; but on examination, we see that, while the Pinks have numerous seeds in a cell, the *Phloxes* have but a single seed. Besides this, the Pinks have a corolla made up of several distinct petals, while the *Phloxes* have but a single or monopetalous corolla, although divided into five deep segments. When Linnæus remodelled botany, he generally retained the old designations if they did not conflict with the requirements of his system, but *Lynchnidea* was one of the names which had to give way. In the first place, the name implied a close relationship to *Lychnis*, which the plant did not have, and thus would mislead. Its form, moreover, was that of an adjective rather than of a substantive, and the system of Linnæus called for an adjective in addition to the substantive. But as *Lychnis* ("lychnos") is the Greek word for "lamp," Linnæus changed the name of the genus to "Phlox," which means "flame," and in this metaphorical way still retained a connection with the original name of Plukenet. The relationship of the *Phloxes* is not with *Caryophyllaceæ*, but with Greek Valerians, with which, and some others, they form the natural order, *Polemoniaceæ*, of which there are numerous representatives in various genera on the American continent. As the *Phloxes* are strikingly different in appearance from most of the plants which were cultivated in Europe at the end of the seventeenth century, their introduction must have been a rare treat to lovers of gardening. The tall forms of *Phlox*, especially *P. Carolina*, found their way into the gardens about 1720; and Peter Collinson, that rare lover of American wild flowers, boasted of several others a few years later. In a letter to John Bartram in 1765, he refers with pride to them, and remarks, "It is wonderful to see the fertility of your country in *Phloxes*." He would have wondered more if he had seen the beauty of the many which have been discovered since his time. Even the one we now illustrate was, probably, unknown to him, as it was not till 1786 that it seems to have been introduced into England by John Frazer.

But beautiful as this species is in gardens, no one can have any conception of its grandeur when seen in some of the wild places where it finds itself perfectly at home. Dr. Darlington, in his *Flora Cestrica*, remarks, "This species is chiefly confined to the Serpentine Rock (in Chester County, Pennsylvania), and when it is in full bloom, the hills, at a distance, apppear as if covered with a sheet of flame." The writer of this has noticed the same lurid appearance of the hills from the flowering plants along the line of the Schuylkill River, as, no doubt, have other observers in other places. It is one of the earliest of all plants to flower in this region. If the autumn be mild, as in Pennsylvania it often is, flowers may be seen as late as in November, while it is not unusual, after a few mild days in the spring, to find some which seem to have opened under the snow, like certain kinds of plants in the Alps of Europe, which, according to Kerner, blossom under similar circumstances.

Many *Phloxes* die completely back to the ground, but this one trails or creeps along on the surface, keeping its leaves as green as moss, and indeed, from this character, has obtained the name of "Moss-Pink." It is also called "Mountain-Pink" and "Ground-Pink"; but the two last names are, perhaps, used only by those "who gather wild flowers," for the commonest garden name is "Moss-Pink." The "Moss" is appropriate enough. "Pink," however, does not properly belong to this genus, but to *Dianthus*, or that family to which the Carnation belongs. It is, doubtless, one of the true Pinks to which Wordsworth refers when he says, —

> "The wild pink crowns the garden wall,
> And with the flowers are intermingled stones,
> Sparry and bright, rough scattering on the hills."

At any rate, Wordsworth's plant is not a *Phlox*, as this does not grow wild in Europe; but our Moss-Pink grows in our country under such similar circumstances, and the flower itself is so like to the real pink of the poet, that the quotation seems to be quite appropriate. All throughout the New England States it delights to grow on rocky hillsides; but as it wanders south, according to Chapman, it takes to low, sandy places. It is found wild in all the States south of New York to Florida, and west to Michigan and Mississippi. In the Rocky Mountains and thence westward, its place is taken by other cæspitose forms which are indirectly allied to it. One species somewhat similar also occurs in Siberia, and this is the only one found outside of the United States.

Writers on medicine have nothing to say about the Moss-Pink, but it has succeeded in attracting the attention of philosophers, for Mr. Darwin gives it a special notice in his "Forms of Flowers." Dr. Gray had noticed that the plant was heterostyled; that is to say, had the pistils in some plants shorter than in others.

In olden times, when these points were not understood as they are now, this short-styled character was thought sufficient to build another species on; and hence Nuttall made one as *P. Hentzii*, in which this was the chief distinction. It shows how great has been the progress of botany even since Nuttall's time, when we see that what are now known to be little more than sexual differences, were then taken to be essential, specific characters. It is this peculiar variation in the length of the pistil that has been noticed by Mr. Darwin. Generally, he found the pollen grains different in size in heterostyled plants; but in this species he found no difference, or, rather, both large and small grains are found on each form, and this he regards as very remarkable. He concludes his notice of this phenomenon by suggesting that "possibly this species was once heterostyled, but is now becoming sub-diœcious, the short-styled plants having been rendered more feminine in nature. This would account for the ovaries having more ovules (two instead of one), and for the variable condition of their pollen grains. Whether the long-styled plants are now changing their nature, as would appear to be the case from the variability of their pollen grains, and are becoming more masculine, I will not pretend to conjecture." Still, the bare suggestion will have an interest to those who are studying what are known as the facts of evolution. Mr. Darwin has evidently a deeper interest in our little plant, in this connection, than his expression, "I will not pretend to conjecture," implies; for in another part of his work, he says, "Certain appearances countenance the belief that the reproductive system of *Phlox subulata* is undergoing a change of some kind." The extracts show with what interest our Moss-Pink is being regarded in science.

To the florist the Moss-Pink offers some attractions. It is not only of easy culture, but is extremely variable in nature, both in color and form. We give some variations on our plate; but there are changes in form as well as in color.

SAXIFRAGA VIRGINIENSIS.

EARLY WHITE SAXIFRAGE.

NATURAL ORDER, SAXIFRAGACEÆ.

SAXIFRAGA VIRGINIENSIS, Michaux. — Low, four to nine inches high; leaves ovate or oval spatulate, narrowed into a broad petiole, crenate-toothed, thickish; flowers in a clustered cyme, which is at length open and loosely panicled; lobes of the nearly free calyx erect, not half the length of the oblong, obtuse white petals; pods two, united merely at the base, divergent, purplish. (Gray's *Manual*. See also Wood's *Class-Book*, Chapman's *Flora of the Southern States, Botany of California*, etc.)

THE names of plants, if literally taken, would often mislead. Michaux, one of our early botanists, finding this plant abundant in Virginia, gave it the distinctive name of *Virginiensis;* but it is distributed over the whole American continent, and is much more common as we go north of Virginia. It is found in Canada and as far south as Georgia, in the Rocky Mountains and in the Sierra and Coast Ranges; and if we accept the suggestion of some botanists that it is scarcely different from *Saxifraga nivalis,* we may say that it runs far away up into the Arctic regions, which is a remarkable geographical range for a plant with no special organs adapted to aid distribution, and to which cultivation and man's work in general are enemies.

The Saxifrages are mostly Alpine or high northern plants, and form a genus of some one hundred and fifty representatives. Only a few of them are found in the Atlantic States, and the species we now describe is perhaps the most southern of all. It is among the earliest in bloom of our wild flowers, being often open in Pennsylvania by the middle of April. It grows in shaded woods or in stony places, and particularly delights in getting into the clefts of rocks. The generic name given to the plant — *Saxifraga* — is from the Latin, signifying "to break a rock," and owes its origin to the fact that some of the species grow in rocky crevices, as we have described this one to do. The common name of the family in Germany is "Stonebreak," but we have become so familiar with the Anglicized Latin Saxifrage that it has entered into our popular botanical language. Our species is known among lovers of wild flowers as the " Early Saxifrage," which, for Pennsylvania and thereabouts, is distinction enough.

It is remarkable that so large and so well-known a family of plants should have proved of so little importance to man. None of the Saxifrages seem to have excited poetic fire, nor have they entered in any way into the arts. Our present species is, however, deserving of some notice for its expressive beauty. Rocks are occasionally met with so rugged and bare that there seems no chance for any living thing beyond mosses and lichens to find a place for existence on them. Scarcely a moss may be seen on their whole surface; yet if there be a ledge or crevice, and it be in the vicinity of the Early Saxifrage, the rock will be found dotted with it. Our specimen was gathered near Germantown, Pa., under just these circumstances there seemed nothing but this plant growing there. In early spring, before the flower-stems have started into growth, there are few prettier sights than a rock sprinkled with these little green plants.

The plant itself affords a good study for the ornamental artist. Before it flowers it forms one of the most beautiful rosettes imaginable. The outline is a perfect circle, and the spoon-like leaves, regularly notched and as regularly disposed around their common centre, give as much variety to the otherwise geometrical form as one can desire, while the little central flower-bud, just ready to push, makes an excellent termination to the whole. For the central ornament in a piece of carving, it would furnish an admirable pattern, or in any case where a starting-point of regular and yet varied form is desirable. As soon as the flower-shoots grow, the lower leaves begin to fade and lose their regular form; but with the warm weather, another attractive feature is developed. The green of the leaves becomes prettily tinted with rose, and at this stage the plant is in nice condition for the artist, to whom these departing shades in the sunset of plant-life are always welcome. The flowers are not showy by any means, being small and colorless; but as soon as the petals begin to fade, the carpels take on a deep shade of brown, which, as we see in our plate, produces a very pretty effect. Many other members of the family have good points similar to those we find in the Early Saxifrage.

Our plant does not do as well on dry rocks as on those on which there is some moisture, and it assumes its handsomest form in shaded places. When the rock has been disintegrated and the remains collect to some depth in favorable places, the Early Saxifrage is in its glory, and will make plants three or four times larger than the one illustrated here.

Some of the species have astringent or aromatic roots, out of which something useful might be made, and in old times one of them was thought to be a good diuretic. None have entered into any of the great scientific questions of the day to any material extent, but they have a use in preparing the bare rocks for better things. The mosses and lichens collect dust on the rocks, and add to this matter by their own decomposition, and the Saxifrages follow, doing much better work after they have once established themselves. In this way, little by little, a surface of earth is accumulated on the rocks; then the rain or melting snow, with the frost, get a chance to operate; and finally, in the course of time, a soil is produced that will grow anything. But this may not be the only service which these plants are capable of doing to man. It is well to note that our knowledge of the uses of things has progressed amazingly of late years, and it is more than probable that this very extensive family still holds secrets which will only be exposed to future generations. Nature does not tell us all she knows at once, but deals it out in small portions at a time.

The Early Saxifrage bears cultivation very well, if not planted in too hot a place, or where the water stands. It can easily be increased by dividing the roots. As of many other species, double forms may also occasionally be found of this. In one of the early volumes of the "Naturalist," such a double form is referred to as having been found in Pennsylvania, and in the volume for 1877 it is noticed that another of the same kind was found. This last is now under cultivation by Mr. Jackson Dawson, the chief gardener of the Arnold Arboretum, at Boston.

ARCTOSTAPHYLOS UVA-URSI.

BEAR-BERRY.

NATURAL ORDER, ERICACEÆ.

ARCTOSTAPHYLOS UVA-URSI, Sprengel. — Corolla ovate and urn-shaped, with a short, revolute, five-toothed limb; stamens ten, included within the corolla; anthers with two reflexed awns on the back near the apex, opening by terminal pores; drupe berry-like, with five to ten seed-like nutlets. *Specific character.* — Trailing; leaves thick and evergreen, obovate or spatulate, entire, smooth; fruit red. (Dr. Gray in *Manual of Botany.* See also Wood's *Class Book;* Watson's *Botany of 40th Parallel;* *Botany of California Geological Survey.*)

THIS pretty spring flower is popularly called the "Bear-Berry." As such it was known all over Northern Europe, where it also grows wild, long before botany was a science and Linnæus, the great botanical "Adam," gave intelligent names to vegetable things. Thus it came that its generic name, *Arctostaphylos*, compounded from the Greek, and signifying "Bear Berry," is derived from the common name, as also is the Latin specific term *Uva-ursi*. It is remarkable that the generic, specific, and common names, though representing three languages, all mean the same thing, — a circumstance that does not often occur in botanical nomenclature. The plant received the name because the bears are said to be fond of the fruit, and the writer of this has had evidence in the mountains of Colorado that this fondness is not a myth. Birds are also fond of the berries, and in Europe especially they are said to be a common food with game. There is no pleasant taste in them to human experience. They are astringent, and this quality gives medical value to them in treating diseases of the kidneys, and where it is desirable to check excessive secretions of mucus. The whole plant, indeed, partakes somewhat of this quality, and is used in the North of

Europe for dyeing gray and black, and for tanning the finer kinds of leather.

The botanical relationship of the Bear-Berry is with the *Arbutus*, from which it is distinguished by having but a single bony seed in a cell. Indeed it was known as *Arbutus Uva-ursi* by the older botanists,—those who may be familiar with the true *Arbutus* will readily recognize the similarity of the flowers,— and it is almost to be regretted that it has not been kept in this genus for the sake of the many poetic associations connected with the *Arbutus Unedo*, which has given the popular character to the family name.

> " Glowing bright
> Beneath the various foliage, wildly spreads
> The arbutus, and rears his scarlet fruit
> Luxuriant mantling o'er the craggy steeps."

This description of the true strawberry tree, "Arbutus," certainly fits our Bear-Berry much better than it does the *Epigæa repens*, to which our people, determined to connect our flora in some way with European memories, have given the name of "Trailing Arbutus," although it has no berry at all. The Bear-Berry has, however, an association with Indian history, as it is the "Kinnikinnick" of the Western races, who smoke it, and believe the practice secures them from malarial fevers. Still, it is almost a pity that the name of "Trailing Arbutus" has been given to the *Epigæa repens*, for, as the Bear-Berry is so nearly an *Arbutus*, and of a perfectly trailing habit, it would be much more applicable to it; but perhaps, if flowers have the affections poets sometimes attribute to them, it was generous in this plant to give up, or rather lay no claim to the name, as whatever might be its own legitimate rights, it is so universally known as the Bear-Berry that it has no great need of the other. The berries are indeed the most striking feature of the plant. The chief resemblance to the real *Arbutus* is in its beautiful white, shining, wax-like flowers. The buds are formed towards the apex of the branchlets in the autumn, and remain in readiness to open as soon as the earliest call of spring is heard. Though the flowers are generally of a smooth, waxy white, they do not seem constantly so, for Mr. Coleman observes that the "corolla and stamens are hairy, in specimens growing at Grand Rapids and other parts of Michigan," and furthermore, that "the margins of the leaves are ciliate, and the petioles and branches pubescent." These facts are very interesting as indicating that, although the plant has so great a geographical range and seems always the same, it may break up in the course of time and form several species.

In its geographical relations there is much to interest the student. Dr. Gray says it is found trailing over rocks and bare hills in the North, and this is, probably, the experience of most collectors. In New Jersey, however, where it is very common, it is generally found growing in sandy pine barrens, and rarely, if at all, on the hills. In Pennsylvania, it grows chiefly along the Delaware, opposite to New Jersey, and in spots that have evidently, from the number of New Jersey plants and the geological character of the soil, been cut off in ancient times from what is now that State, by changes in the river-bed. In the West, it is also found on the sandy shores of the great lakes. On the western side of Lake Michigan, it collects the dry, blowing sands in winter, and the new growth pushes through in spring, in this way increasing in size from year to year, at length forming hillocks of many feet high. The effect in spring, when these hillocks are covered with blossoms, must be very beautiful, and the writer of this can testify to the unique appearance in autumn when the holly-like berries upon them have ripened. The Bear-Berry does not seem to be abundant in Ohio, but has been found by Mr. Beardslee near Sandusky. It is found along the Potomac, and though not referred to by Chapman in his "Southern Flora," is reported from Hillsville, in Virginia, by Dr. Haller, and, no doubt, exists much further South.

The Bear-Berry has the reputation of being opposed to garden culture; but, borrowing a hint from Nature along the lakes, a frame to hold sand was placed around the plant and filled up till only the branch points were left above. Since then it is one of the most luxuriant plants in the writer's garden. To increase the plants, the young stems are drawn up through the hole in the bottom of a flower-pot, the pot filled and sunk in the sand, and suffered to remain without further care for a year or so, when they are separated from the parent and helped to set up for themselves in sand-boxes in the garden.

Mr. E. Hall reports that the plant is very abundant in the coast ranges of hills in Oregon, and is generally diffused through the State. In the Rocky Mountains it is also very abundant, but, according to the writer's own observations, chiefly along the hillsides, where a considerable quantity of disintegrated rock had accumulated.

The fondness of the birds for the berries has, no doubt, aided its distribution, for it is found in tolerable abundance in almost all northern countries, in the language of the "Botany of the Californian Geological Survey," "extending round the world." Though abundant in Oregon, it hardly reaches California, however, where other species replace it.

The flowers, in some European works, are represented as of a rosy pink, but all that we have seen in our country have simply a rosy mouth to the white, waxy corolla, thus really giving it greater beauty than if it were of one uniform tint. Though there are several flowers in one cluster, we have never seen more than one berry mature. Why the remainder are barren is not quite clear. Our drawing was made from a Michigan specimen, in flower on the 26th of April, showing how early it comes into bloom.

TEPHROSIA VIRGINIANA.

VIRGINIAN GOAT'S-RUE; HOARY PEA.

NATURAL ORDER, LEGUMINOSÆ.

TEPHROSIA VIRGINIANA, Persoon.—Erect, villous; leaflets numerous, oblong, mucronate; raceme terminal, subsessile among the leaves; legume falcate, villous; perennial; plant 1 to 2 feet high; stem simple, very leafy; leaflets 15 to 27, 10 to 13 lines by 2 to 3 lines, straight-veined, odd one oblong-obcordate; petiolules one line long; stipules subulate, deciduous; flowers as large as those of the locust, in a short, crowded cluster; calyx very villous; banner white; keel rose-colored; wings red. (Wood's *Class-Book of Botany*. See also Gray's *Manual of the Botany of the Northern United States*, Chapman's *Flora of the Southern United States*, and Torrey & Gray's *Flora of the United States*.)

TO those who live in the vicinity of New York or Philadelphia, New Jersey is a favorite botanical hunting-ground. Our plant may be found there in some abundance in the drier localities during June and July, and it is sure to excite admiration. The color of the flower is not brilliant, but it is sufficient to attract attention, and the neatness of its structure, with the somewhat graceful habit of the foliage, afford pleasure to those who are artistically inclined. The impression the plant gives is one of novelty, for it has more of the character of plants from the Cape of Good Hope, or from Australia, as we see them in green-houses, or judge of them from herbarium specimens, than of those which we generally see in the Atlantic United States. Indeed, species of *Tephrosia* abound in Southern Africa and the East Indies; and speaking of plants as if they had all wandered from a central point, we might say that our *Tephrosias* had really wandered far away from their original home. We have but a few species in the United States, but there are some in the West Indies and in Mexico, and in the southern part of the American continent. It is not by any means in New Jersey only that our plant is found in abundance, for it is frequently met with in wild, uncultivated places from Canada to Florida, west to the Mississippi River, and even beyond, in Arkansas and Texas, to some extent. It varies, however, in some of these districts; so much, indeed, that several species have been made out of it. The leaves change somewhat in these different places, both in form and hairiness, being sometimes nearly smooth. The color of the flowers is also darker in some places than in others. In Michigan, according to Mr. N. Coleman, the two outer petals are almost green.

The silky appearance of the leaves of some of the earliest known species suggested the botanical name *Tephrosia*, "tephros" being Greek for "ashen gray," which is the appearance these silky-haired leaves present; our species exhibits the same characteristic, almost as much so as those which gave the family name. In the time of Linnæus, however, it was not known as *Tephrosia*, but as *Galega Virginiana*, under which name it must be looked for in the earlier botanical works. The original *Galega officinalis* has been left almost alone, the greater part of the many scores of species which once formed that genus being given to its newer-born rival, *Tephrosia*, chiefly on account of their flat pods or seed-vessels, for the original *Galega* has them almost torulose or round. Besides this the vexillum or standard, as the upper petal is called, is longer in *Galega* than in *Tephrosia*.

The separation from *Galega* has deprived our plant of much of its early family history, for *G. officinalis* was the common "Goat's-Rue" of the early writers. Rue itself is another plant, and was used by the old monks to drive away evil spirits that, without proper reason, insisted on bothering mankind. An old writer tells us that these Satanic imps held in utter detestation holy water, Rue, and some other things. The *Galega* was not called Goat's-Rue, however, because it served goats as its namesake served evil spirits (as many persons who want to have gardens where others want goats might well wish), but rather from a slight resemblance in the leaves to the true Rue. The qualifying term, "Goat's," was added because goats eat it with avidity in the places where it grows naturally. In old times the ancient Goat's-Rue was supposed to have strong cordial qualities, and perhaps if it had, the goats, borrowing a hint from a portion of mankind, might have been glad of a little stimulant to a naturally festive disposition. Some of the *Tephrosias* have a very severe character of this sort, and are used to intoxicate fish. The leaves are powdered and thrown into the water, and they act so powerfully on the fish that many of them never recover, but die. This particular species, *T. toxicaria*, is cultivated in the West Indies especially to furnish material for this form of fish hunting. Our plant, *T. Virginiana*, has been found to have some of the virtues ascribed to the original *Galega*. Dr. Wood regarded it as a mild, stimulating tonic and laxative, and used it with good results in typhoid fever. He prepared it by mixing eight ounces of the plant with two of *Rumex acutus*, or, as we now say, *R. oblongifolius*, the Common Field-Dock, in four quarts of water, and boiling the decoction down to a quart; after straining he gave it in doses of one or two tablespoonfuls. When the Europeans came here, they found it a popular vermifuge with the Indians, who used the roots in that capacity, and our people regard it as very useful still. These roots are very long, travelling a great way under ground, and are so tough and wiry that they have procured for the plant the name of "Catgut," under which it is known in the South, in allusion to the similar toughness of violin strings. In most botanical works, however, it has retained its old name of Goat's-Rue (although probably never a goat in America ate it), and as Virginian Goat's-Rue it is often referred to in popular writings. Dr. Gray and some others have used a translation of the botanical name for a common one, namely, Hoary Pea, which is much more characteristic than Goat's-Rue, and worthy of adoption. This would make our species the Virginian Hoary Pea, and all we can say is that our readers have a choice of names. Dr. Peyre Porcher tells us that in the South it is often called "Turkey-Pea."

Though so common in the wild regions of the East, it has not yet found its way into cultivation in our gardens, and, indeed, it does not appear to be in any of the gardens of Europe, although the people there make great efforts to get everything attractive from all parts of the world. It has, no doubt, often been introduced there, but seems impatient of horticultural restraint and gradually pines away. Indeed, an English floricultural writer of sixty years ago says of it, " Though this plant is tolerably hardy in our country, it is nevertheless difficult to preserve it in gardens, for the seeds rarely ripen in England and the plants are often destroyed in winter by the frost." It may be observed that the frost it endures here is more severe than any in Europe, but it is found that many plants which have a high summer heat will endure more cold in winter, and in this way the cooler summer temperature of Europe is not favorable to great endurance in the winter season. In relation to the difficulty of keeping it alive in Europe, Mr. Philip Miller, another celebrated garden-writer of the past age, says, " The only method by which I have been able to keep these plants has been by potting them and placing the pots under a common frame in winter, where they enjoyed the free air in mild weather, but were protected from frost; they have been kept in this way for three years, but never ripened seed in our climate."

Although, as we have said, there are a great many points of interest in the Virginian Hoary Pea, yet the plant is by no means of the highest type of beauty. The thick peduncle, suddenly terminating in the short, thick-set cluster, has a rather " hunchbacked " look, and the gray green is odd, but that is all. The elegance of its leaf-outlines is its redeeming feature. Still it is a plant much more worthy of culture than many which have a place in gardens, and our own florists might perhaps be more successful with it than those of England.

EXPLANATION OF THE PLATE.—1. A flowering branch.—2. Under ground stem, or rhizome.

SEDUM NEVII.

NEVIUS' STONE-CROP.

NATURAL ORDER, CRASSULACEÆ.

SEDUM NEVII, Gray. — Stems low, three to five inches, ascending; leaves alternate, scattered linear-clavate, obtuse; flowers sessile, scattered along the widely spreading or recurved branches of the simple cyme; bracts linear, longer than the flowers; sepals linear-lanceolate, acutish, as long as the lanceolate white petals; stamens eight, shorter than the petals; anthers purplish-brown; carpels tapering into the short, subulate style. (Chapman's *Flora of the Southern United States.* See also Gray's *Manual of the Botany of the Northern United States.*)

SEDUM is a name used by Pliny and other old Roman writers; and Ainsworth and other lexicographers apply it to our common Houseleek, — *Sempervivum tectorum*. The old English writers knew no difference between *Sedum* and *Sempervivum;* and Houseleeks and Stone-Crops, such as we understand by *Sempervivum* and *Sedum*, were mixed together by them, so far as these Latin names are concerned, although they had a separate place in their works for Houseleeks, as distinct from what they thought Stone-Crops to be. This little piece of history is important in connection with the origin of the name *Sedum*, which all our text-books tell us is from *sedeo*, the Latin verb "to sit," and is supposed to have been given to these plants from the habit of growing on bare rocks, sitting, as it were, on them; but we must remember that the name is a very old one, and was merely adopted by Linnæus, because he found it in connection with these plants. We must suppose that there was nothing particularly novel in a plant seeming to sit down, as if it had no roots, for there are many plants which would convey the impression of sitting down quite as vividly, if not more so, than this one, and we shall, therefore, have to look to some more plausible reason for the origin of the name. It seems much more probable that it comes from *sedo*, "to assuage," and has reference to the healing properties of the Houseleek, which latter, as already noted, is regarded as the original *Sedum*. The Houseleek, indeed, has been for ages one of the most popular remedies for relieving pain. An old herbalist says of it, "The leaves, bruised and laid upon the crown or fore part of the head, stop the bleeding of the nose very quickly; and being applied to the temples and forehead, it eases the headache, and allays the beat and distemper of the brain through fevers, frenzies, or want of sleep. The green leaves allay all sorts of inflammations in any part of the body, as in erysipelas or Anthony's fire, and all other hot eruptions of the flesh and skin; and when applied to the sting of nettles or of bees, it presently takes away the pain." Indeed, it is because of its great use in burns and scalds that it is so commonly found in old-fashioned gardens, grown by old-fashioned people, who have more faith in herbs at hand than in the prescriptions of physicians. With this popular impression of the value of the Houseleek, and the connection of the plant with the ancient appellation *Sedum*, it seems probable that *sēdo*, and not *sĕdeo*, is the root of its name; and this becomes still more probable when we note that *e* is used long, and not short, as in *sedentary*, as it would probably be if the two words were really derived from the same root.

The name Stone-Crop may, perhaps, have been derived from the plant "sitting" on stones. The old Saxon word *crop* signified the top of anything, as when we say the "rock crops out," we mean we see the top of the rock above the ground; and as many of the rocks of England are "cropped" with *Sedums*, and cropped by them in a very beautiful manner in many instances, there is no difficulty in accepting this as the origin of the common name.

Many of the old-world *Sedums* have a hot, biting taste, as for instance the *S. acre*, or "Love-entangle" of old gardens, giving, rather than assuaging, burning sensations, as the original name suggests. But no mention is made by any author of any particular qualities, good or bad, in connection with American species. They will, therefore, be of interest chiefly to the lover of the curious in nature and the cultivator of flowers, to whom all the species are very welcome.

Our present species is one of the handsomest of American kinds. It has not long been known, having been discovered within the past twenty years by Dr. R. D. Nevius, a clergyman of Alabama, on rocky cliffs near Tuscaloosa in that State. The botany of the South has not yet been well worked up, and zealous collectors are continually finding new species which have wholly escaped the notice of others before them, or new locations for some that have been supposed rare. Since Dr. Nevius found this plant, Mr. Wm. M. Canby has collected it on Salt Pond Mountain in Virginia, and Mr. Howard Shriver on the rocky banks of the New River, still farther north, and it is quite possible that it may be found abundantly in many other places in the great Alleghany range. In regard to its beauty when growing in its natural location, Mr. Shriver thus spoke of it in the first volume of the *Botanical Gazette*, at the time of his discovery on New River: "Our cliffs are now (early spring) covered with saxifrage, draba, and a variety of *Sedum* with snowy flowers. The stems shoot up from amid rosulate leaves, which are obovate or very short spatulate, often not rounded, but wedge-shaped, giving the idea at first of *Draba ramosissima*. Stem-leaves spatulate to linear spatulate, close set on the high simple stem, and more sparingly on the three branches at the summit. Parts of the flower in fours (the centre one in fives), ovate-lanceolate, somewhat pointed petals twice the length of the ovate blunt sepals. It is probably *S. Nevii*, which Mr. Canby found on Salt Pond Mountain." Our full-face view of an enlarged flower (Fig. 2) accurately illustrates the plant as described by Mr. Shriver, although the specimen from which the drawing was made, and for which we are indebted to Mr. Jackson Dawson, of the Arnold Arboretum, was a cultivated one. It will be observed that Mr. Shriver's description differs a little from Dr. Chapman's, which we have adopted, as to the relative length of the sepals and petals; but as Dr. Gray makes no reference at all to the sepals or petals, it is probable there may be variation in this respect, and these characters may therefore be of no specific importance.

As a general rule, *Sedums* in cultivation like exposed and warm, dry places, but this is true only of the kinds which are found naturally in low regions. Mountain kinds, though they do like open places where there is full light, as a general rule do not like a high temperature. In the effort to cultivate this species made by the writer, it was placed in a piece of rock-work, with a large number of European species, but it gradually dwindled away.

In the culture of *Sedums* we have found that, in spite of their succulence, they seem very grateful when suffered to grow where they can get abundant moisture as well as light. But this moisture must be only for the foliage; for if the roots be in the least stagnated with water, the plant suffers, — rots away in fact. The most successful *Sedums* we ever saw were on a ledge of rock; but they were continually washed by spray from a fountain near by, and thus kept up a beautiful, carpet-like green mass of herbage.

Besides the discrepancy between the characters of the sepals and petals in the plants seen by Mr. Shriver, and those described by Dr. Chapman, there seems to be some variation in the time of flowering. Dr. Gray says the flowers appear "three or four weeks later than" those of *Sedum ternatum*. But Dr Chapman says *S. ternatum* flowers in May and June, and *S. Nevii* in April and May.

In *Nevii* the specific appellation is of course derived from the name of the discoverer of the plant, Dr. *Nevius*, and we may therefore give for its common name "Nevius' Stone-Crop."

EXPLANATION OF THE PLATE. — 1. Complete plant. — 2. Full-face view of an enlarged flower.

PLATANTHERA FIMBRIATA.

GREAT FRINGED ORCHIS.

NATURAL ORDER, ORCHIDACEÆ.

PLATANTHERA FIMBRIATA, R. Br. — Lower leaves oval or oblong, the upper few passing into lanceolate bracts; spike or raceme oblong, loosely flowered; lower sepals ovate, acute; petals oblong, toothed down the sides; divisions of the pendent large lip fan-shaped, more fringed. (Gray's *Manual of the Botany of the Northern United States.* See also Wood's *Class-Book of Botany*, under the name of *P. Bigelovii*.)

THE early American botanists found great difficulty in studying orchideous plants. Muhlenberg, one of the earliest, writing to William Bartram in 1792, says to him that he could never satisfy himself about *Orchis, Ophrys*, and some other genera which he mentions, because, as he continues, "they are so badly described in some, and have too many species in others. I intend to transcribe my descriptions for your perusal and criticism by and by, and hope you will assist me in clearing up some of the rubbish." We in modern times, when we have so large a list of names to choose from, can appreciate the labors and troubles of our forefathers, for in the cases of most Orchids there are long lists of synonymous names which have been given to the same thing by different authors at various times. With the botanists of the past age our plant was *Orchis fimbriata*, and since that time it is *Platanthera fimbriata*, or *Habenaria fimbriata*, according to the different views of authors, setting aside other names not so well known. The modern distinction from the true *Orchis* consists chiefly in the anthers being covered by a pouch in the latter genus, while in *Habenaria* or *Platanthera* they are naked, as we may see in our enlarged Fig. 4. Dr. Gray, to whom we have made our leading reference, unites *Platanthera* with *Habenaria*, as he does not regard the separating characters of full generic importance. The leading difference is that the two glands are approximate in the one section, and widely divergent in the other. There is, however, so little natural distinction between these and several other genera from the true *Orchis*, that most authors have to explain the reasons for the adoption of the several designations, and each branch of the Orchis family may feel a pride in the family history of the ancient name. Our species, indeed, approaches in general aspect many of the European species that have made their mark in the pages of polite literature. Thus in "Hamlet" the queen describes the manner of Ophelia's death, and says,—

> "There is a willow growing o'er a brook,
> That shows his hoar leaves in the glassy stream,
> Near which fantastic garlands she did make
> Of crow-flowers, nettles, daisies, and long purples,"

which last, she says, are also called "Dead Men's Fingers." "Long Purples" and "Dead Men's Fingers" were common names applied to many species of the genus *Orchis* in England, but little different in appearance from the one illustrated here. Our species has not a root quite so characteristic of a "dead man's fingers" as some of the English ones which suggested the name; but as we see in our Fig. 1, there is enough resemblance to claim association with the original idea. Rev. Mr. Ellacomb, writing in Mr. Robinson's "Garden," shows that the name, in allusion to the death-like flesh of the roots, is of great antiquity. He quotes an old ballad as follows:—

> "Then round the meddowes did she walke,
> Catching eache flower by ye stalke,
> Such as within the meddowes grew,
> As Deadman's Thumb and Harebell blew;
> And as she pluckt them, still cried she,
> 'Alas! there's none e'er loved like me.'"

In our country the former botanical name, Orchis, has been adopted in popular parlance; hence this species is known as the "Great Fringed Orchis," and by no other name that we are aware of.

There are many interesting features in a study of this species. The long slender spur which we have endeavored to show in Fig. 6 is characteristic of many species of the genus, and suggested the name *Habenaria*, from *habena*, which is Latin for the round leather leash of a whip, or part of the reins or harness of a horse. *Platanthera*, the generic name, is from the Greek, signifying "broad anthers," from the divergent cells of the anthers, as seen in Figs. 4 and 5. The insectiform look of the flower (Fig. 5) is very interesting; but the most striking feature is seen in the two eyes of a moth or butterfly, which are suggested by the divergent anther-cells. The petals also are peculiarly interesting from their great dissimilarity. The Orchid flower is formed on a ternary type. The lower verticil is composed of a whorl of three sepals, and the upper of three transformed leaves or petals. In our Great Fringed Orchis we see that two of the sepals are nearly equal and opposite to each other, while the third of the series is at right angles with them, and smaller, as we see through the two upper petals. But the three petals, or upper leaves, are still more unequal than the three lower, or sepals; and we see that they have been twisted so that the two conforming ones are on the top, while the more isolated one takes the lower place, and becomes the "lip." In its isolation it has, however, become the largest instead of the smallest, as is the case with the odd one in the lower series, and has so divided itself that it appears as if made of three distinct leaves or petals, and each of these lobes, with its fringed edgings, seems to have a separate existence of its own. The form of the Orchid flowers is so much out of the usual course of nature in flowers, as sometimes to be thought difficult to study, but if we get down to the foundations of the structure, and understand the plan on which the flowers are built, few will be found that are easier. In order that the student may still better perceive the ternary plan on which this Orchid is formed, we give in Fig. 3 a cross section of the ovary. If we examine the stem, we find the leaves scattered thereon. In the stem growth there has been a gradual elongation, but we see that it takes but three leaves to make a full circle round the stem. We do not notice indications of the spiral growth which takes these leaves round the stem, but it is there. It is the more sudden twisting and arresting of the elongating growth that make the set of three sepals and of three petals. These lengthenings and twistings do not go on with regular intensity, but as in waves, sometimes fast and sometimes slow. Such varying intensity and sudden change of degree cannot be seen in many flowers; but this Orchid, as well as some other species, gives a good opportunity for observing it. If we watch the growth of the flower we shall find that it first makes a slow, elongating growth, and that the twisting comes on suddenly, usually taking but a few hours to make a half-turn. In Fig. 2 we have shown an almost mature seed-vessel with the twist towards the base. The torsion in the ovarium is also shown in Fig. 6.

The Great Fringed Orchis seems to have been first made known to botanists through Dr. Pitcairn, who introduced it from Newfoundland to the Kew Gardens in 1777, and it was named *Orchis spectabilis* by Willdenow twenty years later. Its best home is still found to be in the northeastern portion of the United States. It extends westward from New England to Michigan, being quite a common Orchid in the latter State. In Northern and Central Ohio it is also common, but is found more sparingly in the southern part of the State. In Pennsylvania and New York it is found chiefly in the higher districts, becoming scarcer as it reaches lower elevations. In New Jersey, which in many respects is the home of Orchidaceæ, it is found only in the more hilly localities in the western part of the State. Our specimen is from Massachusetts.

EXPLANATION OF PLATE.— 1. Whole plant with root and hollow stem.— 2. Seed-vessel approaching maturity.— 3. Transverse section of the same.— 4. Divergent gland of the anther-cells.— 5. Full-face view of single flower.— 6. Side view of flower.

LIMNANTHEMUM LACUNOSUM.

FLOATING HEART.

NATURAL ORDER, GENTIANACEÆ.

LIMNANTHEMUM LACUNOSUM, Grisebach. — Leaves entire, round heart-shaped, one to two inches broad, thickish; petioles filiform; lobes of the white corolla broadly oval, naked, except the crest-like, yellowish gland at their base, twice the length of the lanceolate calyx-lobes; style none; seeds smooth and even. (Gray's *Manual of the Botany of the Northern United States.* See also Chapman's *Flora of the Southern United States*, and Wood's *Class-Book of Botany.*)

THE plant which forms our present theme affords us an excellent lesson in regard to the meaning of botanical names. When we hear for the first time the name of a human individual, we do not concern ourselves about its meaning in any relation to the person bearing it. We like to know its history, for its own sake. That some Mr. Baker or Mr. Taylor had a primeval ancestor who followed baking or clothes-making may, perhaps, have been the reason why he and all his posterity bear that name; but we do not expect the persons so named now to follow these occupations. A name which means nothing is just as good as one with the most expressive of meanings. Now, many persons think names which are expressive should be given to plants; but expressive names so often mislead that those which have no meaning of any immediate application to the plant in question are generally preferable. For this reason those which commemorate the services of botanists are much in favor with many who describe new plants. How names capable of special application may mislead, is shown in the present instance. *Limnanthemum* is derived from two Greek words, *limne*, mud, and *anthos*, a flower, because, as one would suppose, the original species grew in a marshy or muddy place. But the earliest known species, *L. nymphæoides*, a European plant, grows under water, where the leaves can float on the surface, and does not seem to occur in situations strictly conforming to those alluded to. It is properly an aquatic, and not a marsh plant, as the name would imply, and as Gmelin (author of the "Flora of Siberia") seems to have supposed. Our species was named *L. lacunosum,* from the Latin *lacus*, a lake, by Grisebach, the author of a Flora of the West Indies, from its actual place of growth, and it might be supposed as a corrective of its generic name. But there are in other countries more species that grow in lakes, so we see there is nothing distinctive in either name; and those therefore who might infer it to be so, would be led into serious error.

In old works our plant has to be sought for under the name of *Menyanthes*, or, as it is spelled by Pliny, *Minianthes*. Some writers contend that this name is derived from *mens*, a month, in allusion to its old reputation in certain diseases, or, as Dr. Gray says, from the fact that the flowers last about a month, while those who adopt the Plinian orthography maintain that it comes from the miniate or red-lead color of the flowers. At a later period, it will be found among *Villarsia*, so named from a French botanist, Villar or Villars. Nuttall has it under *Villarsia*, and Michaux and Muhlenberg under *Menyanthes;* but all our modern botanists are united on *Limnanthemum*. It differs from *Menyanthes* particularly in the shape of the corolla, which, when expanded, is wheel-shaped, as seen in our Fig. 4, while that of *Menyanthes* is formed like a funnel.

The flowers proceeding from the petiole or leaf-stalk will, of course, attract attention, and their position will afford a good lesson in vegetable morphology, as showing the intimate relationship between leaves and the axis. It will be remembered by the student that a flower does not consist simply of modified leaves, but of the modified stem and leaves, — a whole branch, and not merely the leaves of a branch. Now, it is a well-known axiom that the lesser cannot be greater than its whole. It follows, therefore, that if the flower, which is essentially stem and leaves, spring from a leaf-stalk, the leaf-stalk must itself possess the same essential elements. Other plants will afford the same lesson in other ways, and we take this one now, simply because the occasion presents itself. Besides the position and nature of the flowers, it will also be interesting to note that roots and buds, making new growths, start out in close neighborhood to the clusters of flowers, so that the petiole or leaf-stalk becomes essentially a stolon, as in the runner of a strawberry, differing from the latter in nothing but its erect position. It is altogether a very good lesson as to how one part of a plant grows out of, or is formed from, another or other parts.

The flowers themselves are very interesting. There are five small sepals, as seen in Fig. 5, and, alternating with them, five petals very prettily fringed and slightly incurved at the edges. (Figs. 4, 3, 6.) Alternate with these, and opposite the sepals, are five stamens, and alternate again with these are five glands. (Fig. 4.) These glands are possibly only another series of stamens, which, by becoming absorbed by the petals in a very early stage, have been aborted. The flowers open and close at regular times of the day, but under exactly what conditions the writer has not been able to determine. The roots remain in the mud during the winter, pushing up in early spring, and by the end of June the flowers appear from underneath the leaf-blades, only a portion of these leaves, however, producing flowers. There appears no difference in strength or vigor between those leaves which flower and those which do not, although there must certainly be a difference in nutrition in favor of the flowering leaves. This, also, is a fact well worthy of remark and further investigation, as in most other plants such a difference in nutrition would manifest itself in a diminished, or increased growth.

It is said by some who have grown certain species of this genus that they are very easy of cultivation, taking care of themselves without any difficulty when once established. This one, however, does not seem to have been taken in charge by gardeners, but would no doubt do as well as any of the rest. For small lakes or ponds it would be very appropriate. The way to plant these, and water plants generally, is to tie them up loosely in thin muslin, with earth and stones, and then sink the whole bundle in the water.

There have been no poetical associations connected with the Floating Heart, as there have been with so many other representatives of the *gentianaceous* order. It seems strange that it has been overlooked. Emblematists might surely have discovered in the dart-like, faded flowers, partly seen from the heart-shaped leaves, some relation to the story of Cupid, and this the more so from its very suggestive common name of Floating Heart.

It is remarkable that there should be but very few species in the genus to which our plant belongs, and yet that there should be representatives of it in every quarter of the globe. Its headquarters seems to be in the East Indies, where there may be half a dozen species. There are, also, one or two in New Holland, about the same number at the Cape of Good Hope, and two in our own country. One is found in Japan, another in Europe and Eastern Asia, one in Brazil and one in South America, with possibly a few others here and there.

Our Floating Heart seems to be abundant in Maine and New England, becoming rare as it reaches New Jersey, although it extends to Florida on this side of the Alleghanies. Its western limit in the north seems to be Ohio, but it travels southwest and is found in abundance in Missouri and Arkansas.

The specimen from which the accompanying drawing was made was kindly furnished to us by Mr. Jackson Dawson, the head gardener of the Arnold Arboretum, Boston, Mass.

EXPLANATION OF THE PLATE.—1. Barren leaves.—2. Fertile leaf.—3. Closed flowers, showing fringed edged petals enlarged.—4. Enlarged expanded flower.—5. Flowers natural size, showing calyx.—6. Flowers, natural size, showing incurved petals.

HOUSTONIA CÆRULEA.

BLUETS.

NATURAL ORDER, RUBIACEÆ (CINCHONACEÆ OF LINDLEY).

HOUSTONIA CÆRULEA, Linnæus. — Glabrous; stems erect, slender, sparingly branched from the base, three to five inches high; leaves oblong-spatulate, one quarter to one third of an inch long; peduncle filiform, erect; corolla with tube much longer than its lobes, or than those of the calyx; flowers light blue, pale lilac, or nearly white, with a yellowish eye. (Gray's *Manual of the Botany of the Northern United States.* See also Wood's *Class-Book of Botany.*)

NO plant is better known than this one in the districts where it grows wild, as it is among the first to bloom in spring, and attracts every one's attention. It was included among the specimens collected in Virginia by Clayton in the last century. The dried specimens which he sent to Gronovius were for the most part described by this celebrated Dutch naturalist, and it was he who named our plant in honor of Dr. William Houston (or Houstoun, as Aiton writes it), an English physician who botanized extensively in Central America, and sent a large number of plants to the Physic Garden at Chelsea, then under the charge of the well-known Miller. Houston was also a contributor to the "Philosophical Transactions," and seems generally to have been a very useful man among the botanists of his time. He died young in 1733; but his friends brought his botanical labors to the notice of Linnæus, who in his earlier works acknowledges his indebtedness to them. Linnæus also adopted the name *Houstonia* from Gronovius, and this will explain why in some works the name is credited to the latter, and in others to the former. It is not customary to go back beyond the works of Linnæus in tracing such records of botanical appellations.

The propriety of the adjective *cærulea* (or *cœrulea*), which means blue, has been questioned by some botanists. Thus an English author of eminence says: "Why *cœrulea*, we cannot tell, for we have never seen any blue about it." This, however, seems rather a strong statement in the face of the combined authority of many other botanists. Prof. Gray says of the little flowers of our species that they are "light blue, pale lilac, or nearly white, with a yellowish eye"; Prof. Wood describes them simply as "pale blue, yellowish at the centre"; and Dr. Chapman, who describes the plant under the name of *Oldenlandia cœrulea*, speaks of them as having the "corolla blue or white, yellow in the throat." It will be seen that blue is given as the leading color by all these authorities; and it may, therefore, be said, perhaps, that the *Houstonia* has as much right to be called *blue* as many another flower.

It is remarkable that so common and so pretty a plant should have remained for so long a time without a generally accepted English name; and yet this was the case, as we learn from Nuttall, who wrote in 1827: "I know no common, prevalent name for our beautiful *Houstonia cœrulea*." About 1830, botanists speak of it as the "Venus' Pride."; and this name still exists, to some extent, in the vicinity of Washington, D. C. In many parts of the country it is termed "Bluets"; but even if we do not object to the association with *blue*, we might ask, in imitation of the English writer above quoted, "Why Bluets?" for the word certainly seems to be altogether meaningless. "Innocence" is also quite a common name, and in some places, according to Darlington and Wood, "Dwarf Pink." Near Philadelphia, the universal name is "Quaker Bonnet," and elsewhere "American Daisy" has also been used. It is rare that we have such an abundance of names to choose from, and one is almost tempted to say that the people, in trying to atone for the long neglect of this modest, yet beautiful little flower, ran to the other extreme of over-naming it. That the first name in the list, Venus' Pride, did not become popular, is hardly to be wondered at; for according to all accounts, Venus was rather a dashing young lady, with a high opinion of her own charms, and such a character is totally at variance with this "wee, modest, crimson-tip'd flower," as our Innocence might be called in imitation of Burns, who in these words characterizes the English daisy. Those poets who have taken this little flower as the emblem of contentment and happiness under poor surroundings have perceived a truth more clearly than is often the case. No flower that we know of so well expresses the virtue of great merits, combined with modesty of bearing, as this. It might well say, with Pope, —

"Honor and shame from no conditions rise ;
Act well your part, there all the honor lies."

But leaving sentiment aside, we find in our plant a great deal to interest those even who care chiefly for material things. Dr. Gray has pointed out that the "flowers are dimorphous in some individuals, with the anthers borne up on the tube of the corolla and projecting from its throat, while the style is short, and the stigma, therefore, included; in the other sort, the anthers are low down in the corolla and the style long, the stigmas, therefore, protruding." Dr. Gray does not notice the additional fact of the dimorphic tubes of the corolla. In the one form, in which the pistil is wholly included, the thick portion of the tube is very short, and the anthers are set on the ledge at the point where the tube narrows (see Fig. 2), while, in the case where the stigma is exserted, the narrow portion of the tube is the shortest (Fig. 3). There is no lengthening of the stamens in either case, but they are simply borne up or down, according to the position of the ledge on which they are placed. In Mr. Darwin's interesting book on "Forms of Flowers," this dimorphism is referred to in connection with some experiments of Prof. J. T. Rothrock on cultivated plants; and Mr. Darwin shows that in the long-styled form the pistil is stronger than in the short-styled one.

There are some facts connected with the distribution of our little Bluet which are also very interesting to the student. While in some districts the plant seems to exist in great profusion, it is sometimes totally absent in contiguous districts in which the circumstances may seem quite as favorable as elsewhere. Willis, in his "Catalogue of the Flora of New Jersey," gives only one locality for it in that State, namely, near Camden; while, on the other side of the Delaware River, it seems everywhere abundant. A correspondent of the "Bulletin of the Torrey Botanical Club," however, says it is also abundant in New Jersey, along the Passaic River, near Newark. The same magazine notices that in the State of New York it may be abundant in some counties, and wanting in others near by. The causes of unequal distribution are worth investigating. With the facts we have given of its irregular distribution, it is not quite clear what its general geographical range may be. Prof. Wood gives it as "found in most grounds, fields, and roadsides, Canada and the United States"; and Dr. Chapman says, "Moist banks, Florida to Mississippi and northward." There is no record, however, of its being found in Michigan, and it is quite likely to be rare in some other States included in the general scope named by the authors above quoted.

EXPLANATION OF THE PLATE. — 1. A complete plant, with barren shoots, half-mature seed-vessel, and flowers. — 2. Half section of a narrow-tubed corolla, showing the stamens near the mouth. — 3. Half section of a thick-tubed flower, with the stamens low down from the mouth.

VIOLA PEDATA.

BIRD'S-FOOT VIOLET.

NATURAL ORDER, VIOLACEÆ.

VIOLA PEDATA, Linnæus. — Nearly smooth; rootstock short and very thick, erect, not scaly; leaves all three to five divided, or the earliest only parted, the lateral divisions two to three parted, all linear or narrowly spatulate, sometimes two to three toothed or cut at the apex; petals beardless; stigma nearly beakless; flowers large, one inch broad, pale or deep lilac-purple, or blue. (Gray's *Manual of the Botany of the Northern United States*. See also Wood's *Class-Book of Botany*, and Chapman's *Flora of the Southern United States*.)

WRITERS have given various accounts of the derivation of the word *Viola*, as applied botanically to the Violets, but most of them rest contented with the simple statement that it is the original Latin name, to which some add, "of uncertain etymology." One of the best modern writers on the Latin language, Ainsworth, considers it, however, to be derived from the Greek. In that language the Violet is called *ion*, and this is a derivative from *ienai*, which signifies "to go." It has been suggested, therefore, that the name was given to our plant from its being a companion to the traveller going through woods and along paths, and in this connection the Latin *Viola* comes to us, *via* being a path or way. This has plausibility to recommend it, and is no worse an explanation than most of those which are offered as solutions of many similar puzzles. It is, at least, pleasant to associate the Violet with wayside travel, for few persons, probably, look back on their childhood, and remember their early rambles along rural paths, without giving the Violet a prominent place in these happy recollections. Whittier truly says: —

> "Not wholly can the heart unlearn
> The lesson of its better hours;
> Nor yet has Time's dull footstep worn
> To common dust that path of flowers."

Our present species, *Viola pedata*, or Bird's-Foot Violet, though we may so pleasantly recall it in the history of our earlier years, is not the earliest to flower when springtime comes. Some few species are ready with their delicate charms as early as the end of March, or by the first week in April, but the "little birdie's foot," as the children pettingly call it, is seldom seen before May. It makes up for its sluggishness, however, by its superior attractions when it does come, for it is the largest and the showiest of all our native species. Not only is it beautiful in its flowers, but its delicately cut and divided leaves give it an elegance which not one of our other species possesses. The Bird's-Foot Violet, also, has a sort of perception of our love of variety, and therefore gives us many forms both of flowers and foliage. This fact is singular enough when we consider it in connection with the statement of a philosophic writer on *English* Violets that, while the pansy, which belongs to the Violet family (*Viola tricolor*), has "bent itself completely to our will, the Violet proper stubbornly refuses to give us any change, and the Violet of the present time is the old Violet of our fathers still." To carry the fancy further, we might say that, hopeless of rivalling the pansy in the affections of the cultivator, the English Violet wisely kept to its own, while our American species, there being no native pansy to compete with it, is trying what it may do to improve. It is remarkable, also, that one of the forms which the *Viola pedata* takes on is not unlike the pansy, as we see by the example given in our plate. This form is by no means uncommon, if we may judge by recent communications in the "Bulletin of the Torrey Botanical Club," of New York, and the writer of this has often had it sent to him as a curious variety by friends in many of the Atlantic United States. In all cases the two upper petals were those that had changed to the beautiful crimson-purple of the pansy; and the reason why, when it does change, it should change in this uniform way, is worthy of the attention of the energetic student. There are similar instances in other plants. Pure white varieties are also very common in some districts, as noted by Prof. Thurber and Dr. I. H. Hale, in the serial from which we have just quoted. In the district from which our illustration was taken, Eastern Pennsylvania, the chief variations are from whitish to purple, and there are many shades between these. But the tendency to vary, far from being confined to the color alone, also manifests itself very markedly in the form of the petals. Some are very broad, giving the flower a round-faced, jolly appearance, while some are mere narrow straps, embodying the thoughtful and careworn expression. There seems to be but little doubt that, in the hands of some enterprising improver, the Bird's-Foot Violet would give highly interesting results. It is remarkable that the English florists, with their known watchfulness, have done nothing in a field so inviting, for the plant has been in their hands since 1759, in which year it was enumerated by Philip Miller as being in the Apothecaries' Garden, at Chelsea, near London, to which it was probably sent by John Bartram, from Philadelphia, with whom Miller commenced exchanging plants in 1755. But perhaps the European florists are so well satisfied with the pansy, that the Bird's-Foot Violet offers no temptation to them. It bears cultivation very well in our gardens, though very seldom seen in the collections of the lovers of hardy border flowers.

Independently of its interest to the mere spectator in the great field of beauty, our plant has also much for those who like to look more closely into the processes of nature. The root, when the plant is taken up, has a bitten-off appearance, or, as botanists say, it is præmorse. Properly speaking, however, this præmorse "root" is nothing but an underground stem,—a little trunk,—and the real roots, thread-like, proceed from it. This stem makes a new addition to its crown every year, and some of the lower portion dies away, just as we see it in the corm of a gladiolus or similar bulb, and this leaves the bottom of the little stem flat, or as if it were bitten off. Indeed, there is actually little essential difference, beyond the shape, between a bulb, a corm, and such a structure as this underground violet-stem. Again, the flower is worthy of close study from its peculiar stigma, which is large, compressed at the sides, and perforated, and very unlike that of most Violets. It is, furthermore, very interesting to study this species in connection with the question of *cleistogamous* flowers, which, as the reader knows, are flowers without petals, fertilized in the bud before the calyx opens, and which follow, during the summer, the complete flowers with petals which cease to appear after June. Nuttall and the earlier botanists believed that all the North American species of Violets produced these apetalous, "secretly fertilized" flowers, but the writer of this has never found them on this species, though he has on most of the others. It is quite likely they may appear in some localities. Large numbers of the flowers give no seeds, but on this and many other points additional observations are much needed.

Most Violets are fond of high elevations, but in the temperate regions some are quite at home when near the level of the sea. Our Bird's-Foot Violet is found in low yet dryish situations, and seems rather to like to get up the hillsides. Mr. Shriver tells us in the "Botanical Gazette," that at Wytheville, in Virginia, it is found in the Alleghanies a half-mile high. Its geographical range commences in Canada and goes down to Florida along the seaboard States, although Dr. Chapman intimates that it has no great love for the warmer parts of the South, but is found chiefly in the upper districts. It extends west to Wisconsin, but is not found in great abundance till it approaches the southern boundaries of the State. In Ohio, Illinois, and Indiana it is frequently met with.

EXPLANATION OF THE PLATE.—1. Part of a root-stock, with leaves and flowers.—2. Bouquet of varieties.

CALLA PALUSTRIS.

BOG-ARUM.

NATURAL ORDER, ARACEÆ (ORONTIACEÆ OF LINDLEY'S VEGETABLE KINGDOM).

CALLA PALUSTRIS, L.—Spathe open and spreading, ovate, persistent; spadix oblong, entirely covered with flowers; the lower perfect and hexandrous, the upper often of stamens only; floral envelopes none; filaments slender; anthers two-celled, opening lengthwise; ovary one-celled, with five to nine anatropous ovules; stigma almost sessile; berries (red) distinct, few-seeded; seeds with a conspicuous raphe, and an embryo nearly the length of the hard albumen. (Gray's *Manual of the Botany of the Northern United States*. See also Wood's *Class-Book of Botany*.)

THE derivation of the name *Calla* is uncertain. Prof. Wood and others believe it is from a Greek word which signifies "beautiful"; but though many of the *Aroid* order are interesting, there are none so striking for their beauty as to suggest a name specially based on that quality. Dr. Gray seems to be of the same opinion, as he confines himself to saying that *Calla* is "an ancient name of unknown meaning." Some of the plants comprised in the genus were certainly known by this name in very remote times; and Dalechamp, a French author of many years ago, believed it was already applied to a species belonging to this family by the ancient writer Pliny. Linnæus, finding it in use in connection with this plant, adopted it as it now stands. If we felt inclined to hazard a mere guess ourselves, we might perhaps say that, inasmuch as most of the species likely to have been known to the ancients are of a peculiar tint of green, the name probably originated in a word denoting a sea-green color.

The *Calla palustris* is extremely interesting, in studying the natural orders of plants, as affording a good lesson on the uncertainty of characters derived from mere sexual distinctions. As noted in our botanical references at the head of this article, Dr. Lindley classes our plant in the natural order *Orontiaceæ*, which was divided from the true *Arums*, as they were then considered, by R. Brown. In support of this arrangement, Dr. Lindley says: " The greater part of these plants (*Orontiads*) have the habit of *Arads*, with which they are usually associated, and from which, in fact, they differ only in having hermaphrodite flowers, which have usually a scaly perianth." But as we see by the description we have given from Dr. Gray, our plant often has the upper flowers staminate only, and there is, therefore, no morphological reason why all the flowers might not be so under some circumstances. In like manner, we have in our plant an absence of the perianth, which, Dr. Lindley remarks, should "usually exist" in the order. These and other considerations fully justify American botanists in not recognizing *Orontiaceæ* as a *natural* order.

The resemblance, in general appearance, of our plant to the common Calla, or *Richardia Æthiopica* of our gardens, is very striking; and indeed the two were for a long time associated together under the same family name. But the Egyptian plant has been separated by Kunth, under the name of *Richardia*, because the anthers have no filaments, — are sessile, — and because of a difference in the cell-divisions of the ovary. Stress is also laid on the fact that, while in *Richardia* the spathe is convolute, and folds around the spadix as a perianth would do in an ordinary flower, in the true *Calla* it is flattened and exposes the spadix to full view.

It is quite remarkable that so pretty a native plant has not found its way into general culture; for though not so striking as its sister, the *Richardia*, or Calla Lily, it has the great advantage of being thoroughly hardy, while the other is destroyed by a very little frost. It seems to be more appreciated in England than here; for Mr. Robinson, in his work on "Alpine Flowers" cultivated in English gardens, pays it a high compliment. He says: "More beauty (in an Alpine garden) than any native plant affords, results from planting in boggy places this small, trailing *Arad*, which has pretty little spathes of the color of those of its relative, the Ethiopian Lily. It is thoroughly hardy, and though often grown in water, likes a moist bog much better. In a bog or muddy place, shaded by trees to some extent, it will grow larger in flower and leaf than in water, though it is quite at home when fully exposed. In a bog carpeted by the dark-green leaves of this plant, the effect is very pleasing, as its white flowers crop up here and there along each rhizome, just raised above the leaves. Those having natural bogs would find it a very interesting plant to introduce to them; and for the moist, spongy spots near the rock garden, or by the side of a rill, it is one of the best things that can be used." We may add that those who have no moist places on their grounds can cultivate this and similar plants by filling small kegs with earth and sinking them in the ground to their rim. As the water cannot readily escape, a sort of a natural bog results, which suits these plants very well in the stead of their natural habitats.

The Bog-Arum is not only a native of the United States, but is also common in Northeastern Europe, and its hardiness may be well understood from its being a very common plant in Lapland. In some of these high northern regions, it seems, indeed, to grow with more luxuriance than it ever reaches in our country. An old writer speaks of it as, in these high latitudes, "growing so vigorously as often to exclude other plants, and occupy whole marshes alone by themselves. They have a hot, biting taste, and yet bread is made from the roots." An English writer of several centuries ago also speaks of an "Arom known as Starchwort"; and it is quite likely that the species native with us is the one alluded to; for Dr. Lindley says: "The rhizomes of *Calla palustris*, although acrid and caustic to the highest degree, are, according to Linnæus, made into a kind of bread in high estimation in Lapland. This is performed by drying and grinding the roots, afterwards boiling and macerating them till they are deprived of their acrimony, when they are baked like other farinaceous substances. It is called *missebræd* in Lapland. The plant has the reputation of being a very active diaphoretic."

Besides in Lapland, it is also reported as being very abundant in Norway and Sweden, Holland, Germany, and Russia, to Siberia. In our own country, Dr. Gray records it as being found in "cold bogs, New England to Pennsylvania, Wisconsin, and common northwards"; and Prof. Wood, as "in shallow waters, Pennsylvania to New England, Wisconsin and British America." Prof. Porter records it as being gathered by him in Northwestern New Jersey. The "Bulletin of the Torrey Botanical Club," of New York, gives, as special locations, " New Durham Swamp," and " Orange County, New York." It seems rather common in Wisconsin, and was found in the northern part of the State of Ohio by Mr. Beardslee. All the leading authors seem to make Pennsylvania its southern limit, but it is included in old lists of the flora of the District of Columbia, though not in the catalogue of the modern " Potomac Naturalists' Field Club." It has not been the writer's privilege to find it wild anywhere himself, and the specimen from which the accompanying drawing was made was gathered in the neighborhood of Boston by Mr. Jackson Dawson.

The specific name *palustris* is, of course, in reference to the marshy places in which the plant grows. Its common name in England, according to Mr. Robinson, is " Bog-Arum." Dr. Gray gives the common name in New England as " Water-Arum." As we have to choose between the two, and Mr. Robinson says it grows better in wet land than in water, we have placed " Bog-Arum " at the head of our description.

EXPLANATION OF THE PLATE.—1. Rhizome and complete plant.—2. Scape, with fruit approaching maturity.—3. Single flower, with stamens and ovary magnified.—4. Cross section of the ovary, showing portion of the ovules.

EUPHORBIA COROLLATA.

FLOWERING SPURGE.

NATURAL ORDER, EUPHORBIACEÆ.

EUPHORBIA COROLLATA, L.—Erect; cauline and floral leaves oblong, narrow, obtuse; glands of the involucre obovate, petaloid; umbel five-rayed, rays two or three times di- or trichotomous; stem slender, erect, one to two feet high, generally simple and smooth; leaves one to two inches long, often quite linear, very entire, scattered on the stem, verticillate, and opposite in the umbel; the umbel is generally quite regularly subdivided; corolla-like involucre large, white, showy. (Wood's *Class-Book of Botany*. See also Gray's *Manual of the Botany of the Northern United States*, and Chapman's *Flora of the Southern United States*.)

DURING the wars between Cæsar and Pompey, one of the partisans of the latter, King Juba, of Mauritania, or Southern Africa, distinguished himself by his martial skill, and has, therefore, had his deeds handed down for the edification of posterity, although finally he suffered a disastrous defeat. The same king is also celebrated in history as being the father of a son, great in science and general intelligence, who bore his own name. But he must himself have been a man of some penetration, if history can be trusted to tell the truth about kings; for it is said that Juba, although he had a very famous physician, himself discovered wonderful medical virtues in a plant growing wild in his dominions. It is furthermore stated that he named this plant after his physician, who was called Euphorbus, and hence our botanical name *Euphorbia*. What particular species it was that the Mauritanian prince thus honored with his attention has not been definitely decided; for in that king's old dominions the *Euphorbiaceæ* abound as thick, heavy, succulent bushes, many indeed being small trees of twenty feet or more in height. These plants have much the appearance of the cactuses of our own continent, bearing spines on the angles of the stems, as our cactuses do, but differing from them in having a milky juice which runs freely on the slightest puncture. In these histories of botanical names, such as the one just related, we must accept the accounts as they are handed down to us, without much questioning. Otherwise, if we were to examine them critically, we might frequently be led to reject them altogether. In the present case, for instance, it might be said that such very common and peculiarly striking plants must certainly have had some recognized name long before King Juba deigned to take notice of one of them. Even our wild Indians give common names to striking plants, and as the literal meaning of *Euphorbia* is "well fed," it might be argued that it is a designation very likely to occur to any one in connection with such fat-looking, milk-gorged vegetation, without necessitating the intervention of a royal intellect.

The common name of the family is "Spurge," and seems to come from the French "Espurge." It is the same in effect as our word *purge*, which expresses the peculiar virtues said to have been discovered by King Juba. All the members of the genus *Euphorbia* possess more or less of this purging character, and a plant of the same natural order, *Ricinus communis*, is indeed the veritable castor-oil plant. Aside from this purging character, the Euphorbias, all of which are poisonous, seem to have no qualities useful to mankind. Our own famous botanist, Nuttall, appears to have had quite a dislike to them, for he speaks of them in a manner unusual in one who always showed so great a devotion to nature in every form. He says: "The economy of the genus *Euphorbia* appears to be very limited. In the deserts of Africa they only tend, as it were, to augment the surrounding scenes of desolation; leafless, bitter, thorny, and poisonous, they seem to deny food to every animated being. Among the European and American species, there are some which have been used medicinally, but they are, at best, dangerous and needless remedies." His contemporary, Rafinesque, however, seems to have had more charity for them, as he selects our plant, the *Euphorbia corollata*, as a leading representative of the "Medical Flora of the United States," giving an illustration of it in the curious work bearing that title. He says that, as a purge, the *E. corollata* is the most efficient of all the American species, as only about three to ten grains need to be taken, and that a dose of from ten to twenty is a good emetic. He further says that the action is always proportionate to the quantity taken, which is not the case with common ipecac, and that it is, therefore, more "manageable and safe." It appears that the peculiar medicinal character of this plant was known to the Indians, and Rafinesque notes as a very singular circumstance the close resemblance of the name given to it by the Indians of Louisiana, " Peheca," to the Brazilian name, " Ipeca," more especially as both words have the same meaning, namely, " Emetic Root."

The root is somewhat fusiform in shape, with very little tendency to branch, and has only a few fibres attached to the lower end. It is covered with a thick bark, which in old plants sometimes constitutes two thirds of the whole root, and in this bark the medicinal properties chiefly reside. It seems to have served a good purpose to the Southern Confederacy during the civil war. Dr. F. R. Porcher, who was one of the medical officers in the Confederate service, says, on the authority of Dr. Frost, Professor of Materia Medica in South Carolina Medical College, who had used it in his practice "with great benefit," that "it is as active as ipecacuanha, and fully entitled to the consideration of the profession. . . . Even should it not be employed, every physician should be instructed in its properties, and, when the occasion requires it, know the substitute he can use in case of need." We have been so particular in recording the opinion of our physicians on this subject, because we were unwilling that so pretty a native plant should be regarded as an utterly worthless thing.

As a matter of scientific accuracy, we must also note, for the benefit of the lover of wild flowers, that the pretty blossoms which he admires are not flowers at all; that is to say, the white structures are not petals, as in ordinary flowers, but merely bracts. It is some comfort to know that the great Linnæus thought they were true perianths, and that he placed the plant in his sexual class, Enneandria, as a single flower, having nine stamens. But really each stamen represents a single flower, as a close examination will show. The stamens come out from the axils of little leaves or bracts, each one having its little home to itself. The female flower is simply an ovary on a short stalk, and occupies the central place in this curious specimen of inflorescence. The flowers are, therefore, monœcious, or in other words, the male and female flowers are separate, although the petal-like semblance of the involucral bracts imparts to the whole the appearance of a seemingly regular hermaphrodite flower. This leafy or bract-like character of these appendages may be better understood by examining the common green-house *Poinsetta*, from Mexico, the scarlet bracts of which are so often found among cut flowers.

The *Euphorbia corollata*, or Flowering Spurge, is widely diffused over the eastern part of the United States, growing (sometimes low and spreading, according to Gray, in his "Field, Forest, and Garden Botany") in open, waste woodlands, and often in badly cultivated fields. It seems to have its northeasterly limit in New York, whence it extends across the continent to Nebraska, down to Arkansas, and from there eastward to Florida, thus making a home for itself over a vast extent of territory. We know of no attempt to cultivate it, not even in England, where so much enterprise is shown in getting together pretty flowering things. It is generally in bloom in July and August, and makes a branching stem about two feet high. Our plate, it will be seen, represents only a portion of the panicle.

ERRATUM.

Part VI, p. 92, line 14 from below, instead of *Orchis spectabilis*, read *Orchis fimbriata*.

POTENTILLA FRUTICOSA.

SHRUBBY CINQUE-FOIL.

NATURAL ORDER, ROSACEÆ.

POTENTILLA FRUTICOSA, L.— Stem erect, shrubby, two to four feet high, very much branched; leaves pinnate; leaflets five to seven, closely crowded, oblong-lanceolate, entire, silky, especially beneath; stipules scale-like; flowers numerous, yellow, terminating the branchlets. (Gray's *Manual of the Botany of the Northern United States.* See also Wood's *Class-Book of Botany.*)

ACCORDING to Dr. Gray, the name of the genus *Potentilla* is "a kind of diminutive from *potens*, powerful, alluding to the reputed medicinal power, of which, in fact, these plants possess very little, being merely mild astringents, like the rest of the tribe." Almost every common plant had some great virtue attached to it by the people of the olden time, and for this one it was claimed that "it is good against all sorts of agues and fevers, whether Continent, Continual, or Contermitting: whether they be burning fevers only, Malign or Pestilential. It cools and attemperates the blood, and Humors, and is an excellent thing for a Lotion, Injection, Gargle, and the like, for Sore Mouths, Ulcers, Cankers, and other corrupt, foul, and running Sores. The juice mixt with a little Honey, prevails against Hoarseness, as also the Cough of the Lungs." These are some of the reputed powers to which Dr. Gray refers, and which suggested the present botanical name of the family. In old writers we find the appellations *Pentaphyllum* and *Quinquefolium*, Greek and Latin names, respectively, for "five-leaved," the leaves of most of the species being in fives, and the present common name, "Cinque-foil," is, of course, identical with these. But the orthography of the latter is French, and it is a matter of surprise that a plant, so common in England as is the Cinquefoil, in numerous forms, should yet seem to have had no distinctively English name whatever.

Most of the *Potentillas*, or Cinque-foils, are creeping plants, or herbaceous plants, with evergreen foliage, such as is the strawberry plant, to which family, indeed, the Cinque-foils are closely allied; but the *Potentilla fruticosa* takes on a woody character, and becomes a small bush, and in this is an exception to all the rest of the family, of which there are nearly a hundred species. Some botanists have, indeed, tried to make several species out of the one now under discussion. In Europe, where it also grows wild, it has long been known; and when Pursh came to this country, in the beginning of the present century, and found the plant here, he believed it to be distinct, and named it *Potentilla floribunda*. Nestler also thought the Russian plant distinct from the general European form, and called it *P. davurica*. Schlechtendal again names a kind with narrow leaves *P. tenuifolia*. But the best authors in Europe, and all in America, agree in considering all these forms as mere varieties of our present *P. fruticosa*.

The student will notice, on examining the circuit of the leaves round the stem of *Potentilla*, that five leaves form a complete circuit, or a verticil, and he will perceive the operation of the same law in the formation of the flower, which is, indeed, nothing but a suddenly arrested branch, the petals and sepals being transformed leaves. He therefore finds a double row of sepals of five each, and five petals in the flower, and the stamens generally some multiple of five. When any of the number is wanting in these cases, it is generally because the convolving and depressing growth has been so rapid as to entirely obliterate some of the petals, or in botanical language, because they have disappeared by abortion. The gradual retardation of the wave growth is very prettily illustrated here. Although most Cinque-foils have but five leaflets, the Shrubby Cinque-foil has often seven; but when growth-force is about to be arrested by reproductive force, only five are formed, and then, successively, only three, two, and one. Thus it appears that the rapid convolutions, which end in the verticils forming the flowers, occur only when the growth-force has been reduced to the production of single leaflets instead of full leaves. If the same thing were to occur before, at the three or five leafleted condition, the probability is that the petals would each be three or five lobed instead of entire, as we see them now. There is also some special interest in the calyx, which, as we have said, is composed of a double verticil of five leaves each. The outer set remains somewhat spreading, but the inner is bent inwards, making a slight covering for the naked seeds (Fig. 3). The result is a very pretty design for ornamental work, as shown in our full-face view of the capsule in Fig. 2. The seeds in this species of Cinque-foil have likewise a special interest of their own. In some of the allied *Potentillas*, the styles are thickened upwards, being what is technically called "clavate" or "club-shaped"; but our species, with a few others, has them filiform, so that, after the petals have fallen, the seeds look as if they were covered by a growth of thin hair. On this account, Torrey and Gray grouped these species together in a separate subdivision, with the expressive name, *Comocarpa*, — *coma* signifying a head of hair.

Potentilla fruticosa is also interesting from a geographical point of view. It is widely diffused over the northern regions; and if we allow the several forms alluded to above to be simply varieties of the same species, we may say that it makes a circuit completely round the globe. It is abundant in Maine, Massachusetts, and Connecticut, decreasing in extent through New York till it reaches a southern limit in Northwestern New Jersey. We know of no locality where it is wild in Pennsylvania, although not uncommon there in half-cultivated places. In its New England locations, it seems to prefer low, wet meadows. In Ohio, it is found in dryer situations. When it reaches Michigan, it loves to grow among the sand on the lake shores; but as it travels farther into the State, it is found on dry, rocky places in the dells. In Colorado, it grows in extremely dry localities, both in the foot-hills and high up in the mountains, and it continues in this way to vary its conditions until it reaches California, where, according to the geological survey of that State, it is found in Ebbett's Pass, in the Sierra Nevada, and thence takes its march northward to Siberia. In Wyoming, Dr. C. C. Parry tells us that, with a few other *rosaceous* plants, it forms almost all the shrubbery they have in that treeless region; but it is only a small shrub, rarely exceeding two feet high in our gardens, where it is very easily grown and very welcome on account of its profusion of bloom from July till October, and at a season of the year when few other shrubs give us any flowers.

In some parts of Connecticut, it has found the soil and climate so much to its liking that it takes complete possession of the ground, to the great annoyance of the agriculturist. It is called " Hard Hack " in those parts; but as this name is better known in connection with *Spiræa tomentosa*, there is no reason why it should supersede Shrubby Cinque-foil. Dr. I. H. Hall, however, in the "Bulletin of the Torrey Botanical Club," Vol. I, says that it is the *P. arguta* which the people of Connecticut call "Hard Hack," and which is so bad a weed there.

It is said to be a remarkable fact that, although all other animals will eat *Potentilla fruticosa* greedily, hogs cannot be persuaded, under any circumstances, to touch it. We have not been able to verify this from experience, and so give it as part of existing history, subject to future experiment; for in these matters repetition of observations does no harm. In some parts of Europe, brooms are made of the branches, which are said to be equal to heath or birch, but the plant has no known use in this country.

EXPLANATION OF THE PLATE. — 1. A flowering branch. — 2. Calyx in full-face view, showing its beauty for ornamental designs. — 3. Calyx, showing the five inflexed, upper sepals.

LINUM PERENNE.

PERENNIAL FLAX.

NATURAL ORDER, LINACEÆ.

LINUM PERENNE, L.—Smooth and glaucous, one to two and a half feet high, branching above, leafy; leaves linear to linear-lanceolate, three to eighteen lines long, acute; stipular glands none; flowers large, blue, in few-flowered corymbs, or scattered on the leafy branches on slender pedicels; sepals three to five nerved, ovate, acute, or obtuse, one and a half to two and a half lines long; capsule globose, acute, exceeding the sepals, at length dehiscent by ten valves, the prominent false partition long-ciliate; fruiting pedicels erect or deflexed. (*Botany of California*. See also Porter's *Flora of Colorado*, Watson's *Botany of the 40th Parallel*, and Wood's *Class-Book of Botany*.)

> "Oh, the goodly flax-flower!
> It groweth on the hill,
> And be the breeze awake or sleep,
> It never standeth still.
> It seemeth all astir with life,
> As if it loved to thrive,
> As if it had a merry heart
> Within its stem alive!"

THE full force of these lines of Mary Howitt never impressed itself so strongly on the writer as when, high up "the hill" in the Rocky Mountains, he gathered for the first time a wild specimen of the plant now illustrated. It was in a particularly barren spot, where even the few things that grow in this inhospitable region hardly dared to risk themselves; but the *Linum perenne* was doing beautifully, expanding its large, blue flowers to the morning sun "as if it loved to thrive" even in so dreary a place. It is found in quite low elevations, but increases in abundance as it travels up the hillsides. The expression that "it never standeth still" applies better to our Flax than to the closely allied European species of *Linum usita-*

tissimum, or "most useful" Flax, of which Mrs. Howitt wrote, and which is an annual, dying after the seed has ripened, while ours is a perennial species, the plant continuing on from year to year. Its continuous growth is, indeed, remarkable. In the early spring we find it little more than a small tuft of green leaves, but it soon throws up from each bud a flower-shoot which by May is covered with blossoms. It does not commence to bloom till it has made its full length, and then the uppermost flower opens first. After this the lateral ones open continuously from the side branches downwards. Those branches which flower first naturally mature first. By September the flowering stems have nearly all ripened, and commenced to turn brown. Other branches, however, still continue to push out from the lower buds on the main shoots; but as if they had an instinctive knowledge that there would not be time to ripen seed before the winter sets in, they make no attempt to flower. These late-growing shoots are just as vigorous as those which, in the early part of the season, threw up flower-stems, and their office seems to be to elaborate sap, and store up nourishment in the crown for next year's floral growth. It is, no doubt, this autumn crop of growth which is the real agent in making our Flax a perennial, while the closely allied European species is an annual. If the latter had its flower-stalks cropped so as to force it to throw out a late, leafy growth below, it would, probably, be as perennial as the American species, and still more "astir with life" than the poetess describes it. The plant in the writer's garden, brought many years ago from Colorado, and from which our drawing was made, is one of the most interesting in the collection, in early winter, by the mass of living green shoots pushing up so freely among the mature and dry stems.

These seed-bearing branches of our Perennial Flax have assumed a new interest since the writings of Mr. Darwin appeared. He finds that some of the flowers of this species have styles longer, and others shorter, than the stamens, and that only the pollen of one plant carried to the flowers of the other plant will enable it to perfect seed. Mr. Darwin says the two forms of stamens "stand at different heights, so that the pollen from the anthers of the longer stamens will adhere to one part of an insect's body, and will afterwards be brushed off by the rough stigmas of the longer pistils, whilst pollen from the anthers of the shorter stamens will adhere to a different part of the insect's body, and will afterwards be brushed off by the stigmas of the shorter pistils, and this is what is required for the legitimate fertilization of both forms. We know that its own pollen is as powerless on the stigma as so much inorganic dust." ("Different Forms of Flowers," 1877, p. 98.) The plant from which we made our illustration has, however, been growing separately and alone from 1873 to 1878, and has no opportunity to receive pollen from other plants, but it nevertheless produces seeds in tolerable abundance every year. This shows that, while in England only cross-fertilization will produce seed, climatal influences bring about different results in America, and the whole indicates that much more remains to be discovered about the habits of plants, and their "sources of action," than has yet been found out. Dr. Gray thinks the American Perennial Flax may not be heterostyled as the Asiatic form is, and may, therefore, be a distinct species.

The Perennial Flax affords much interest in its flowering. The young tips of the flower-shoots droop down. When the buds are ready to expand, they assume a perpendicular position during the night, and by morning the flowers open, turning towards the rising sun. Long before noon the petals have performed their functions and have withered away. Mr. Darwin has noticed a peculiar twisting of the pistils, which places the stigmatic surface towards the circumference of the flower. This, however, he finds confined to the long-styled forms. No doubt many more discoveries of interest would reward careful observers of the behavior of this plant.

The specific name *perenne* indicates the most striking distinction between our species, and the one which yields the ordinary Flax. This, however, is not all the distinction, nor would it be regarded as in itself sufficient for botanical science to build on. as, in the present condition of botanical knowledge, so much importance is not attached to slight variations as there was in old times. The native country of the common Flax, *Linum usitatissimum*, is not known, and it is not at all improbable that it is only a form selected and used for cultivation. Flax has been grown for ages for its fibre, of which fine linen fabrics are made; and in the twelfth chapter of Genesis, we read that Pharaoh clothed Joseph in fine linen; and again, in the fourth chapter of Exodus, that, when the plagues came on the Egyptians, the smiting of the Flax crops was one of them. The plant mentioned in the Bible was formerly supposed to be identical with the common Flax; but seed-vessels found in old bricks and similar material from ancient Egypt show that the Egyptian Flax was not the *L. usitatissimum*, but rather *L. angustifolium*, which is also a perennial species, and scarcely, if at all, different from our *L. perenne*. There is, besides, another perennial form, native to Eastern Asia, the *L. perenne Sibiricum*, also scarcely different; and all this renders it highly probable that the true Flax is a descendant of our species. An additional proof that it may have had this origin is the fact that the common Flax varies remarkably in itself. At the American Centennial Exhibition in Philadelphia, a great number of varieties came from Russia and Holland, differing as much among themselves as the whole, as a species differs from our perennial Flax. The probable close connection of our plant with the linen of the mummies and the literature of the ancient people will give our plant a new interest in the eyes of the lover of American wild flowers.

Our plant seems first to make its appearance near the Mexican boundary, whence it traverses the whole continent between the Pacific and the Mississippi, extending through its several varieties to Europe and Asia.

XANTHOSOMA SAGITTIFOLIA.

ARROW-LEAVED SPOONFLOWER.

NATURAL ORDER, ARACEÆ.

XANTHOSOMA SAGITTIFOLIA, Schott. — Stemless; leaves glaucous, hastate-cordate, acuminate, the lobes oblong, obtuse; spathe hooded at the summit, oval-lanceolate, white, longer than the spadix; root tuberous; petioles twelve to fifteen inches long; leaves five to seven inches long, the lobes somewhat spreading and generally obtuse; scape as long as the petioles. (See Chapman's *Flora of the Southern United States*.)

VERY few persons who go out to gather wild flowers will return with the subject of the present sketch, for it is one of the scarcest of our native plants. The writer has never met with it in a wild condition, and the drawing was made from a specimen kindly furnished by Prof. C. S. Sargent, of the Cambridge Botanical Garden, Massachusetts. Dr. Chapman, whose description is here adopted, gives only two localities, Savannah, Ga., on the authority of Elliott, who was the author of an early botany of South Carolina; and Wilmington, S. C., on the authority of Dr. Curtis. It is scarcely likely to be confined to these two places; but if any other botanist has collected it elsewhere, it is in no list at our command. It is not at all unlikely to be found in Florida, and perhaps many other places South; for these districts have not yet been very well explored botanically. But however that may be, *Xanthosoma sagittifolia* is certainly not a plant which had its home originally in the United States, though it may have been on our soil for countless ages. It is more probable, on the contrary, that it is a wanderer far away from the original centre of its primeval being. It is very abundant in the West India Islands, which, in almost all European botanical works, are mentioned as its only place of nativity; while it is but rarely thought of by any European writer in connection with the United States. It is also believed to be a native of China, where it is extensively cultivated.

Our plant is quite closely allied to a very common garden plant,—*Caladium esculentum*,— the "Tanyah" of the Southern States; and the tuberous roots of both are of equal value. In a raw state, the roots of *Xanthosoma*, like those of most of the *Araceæ*, are extremely acrid, and blister the mouth when brought into contact with it; but this acridity is driven out by heat, and when the roots are cooked, they are very mealy and agreeable, and said to be almost precisely like those of *Caladium*. In China, we are told, the leaves also are used, cut and boiled like our spinach, and they are said to be an excellent vegetable when prepared in this way. It may be well to observe, for the benefit of those who may desire to cultivate the plant in our country for culinary purposes, that, judging by its probable central home, it is not likely to endure any frost; but the roots can, no doubt, be preserved in the winter in any dry place where the thermometer does not fall much below forty-five degrees, although, like the *Caladium*, it seems naturally to be at home in wet, marshy, or springy places.

To most of our readers, however, the edibility of our plant will be but an incident. Its chief interest will be in its beauty, and the botanical lessons which it affords. The resemblance between it and the common Calla Lily, *Richardia Æthiopica*, of our gardens, is seen at a glance, and it gives the general appearance of being something between that and the *Caladium* before referred to. The last named has the flowers low down, scarcely rising above the bulb; while the Calla Lily sends them above the leaves. Our plant has them about of equal height with the leaves, nearly in the position shown in the plate. The shape of the leaves, and at the first glance also the flowers, remind us strongly of our common garden plant. Indeed, the differences in most of the genera of

Araceæ are founded on characters that relate to sexual peculiarities, and are open to about the same objections as the sexual system of Linnæus, which prevailed before the present natural system of botany was introduced. Under the old sexual system our plant would have been associated with an Orchid, or with, perhaps, even a Papaw (*Asimina triloba*), all on account of the peculiar relations of stamens with pistils. Now those plants which are alike in general characters are brought together, and the order which results — *Araceæ* or the *Arum* family in this case — is a very natural-looking one, which the youngest student can scarcely fail to recognize; but when we come to divide the order into genera, we have still to take into consideration the sexual relations; and the result is that we can hardly tell, when we examine a plant of the order, in which genus to place it. The spadix — the central body — has the flowers variously arranged over its surface, and this is regarded as a matter of great importance in determining the genus. In some the spadix is quite naked at its end; in others it is clothed, generally to the apex, and here we find one great difference between the Calla Lily and our plant, for the latter would be placed in the first section, while the Calla belongs to the last. The differences in structure, and the relations of the anther with the connective, are also taken into consideration in determining the genera. In one great division, in which we find the true *Arums* and our Indian Turnip, the cells of the anthers are larger than the connective; in another in which our plant is found, they have a very thick connective; while in the section which contains the *Richardia*, they are embedded in the connective, which is very thick and fleshy. We see by this that plants, which must be closely allied from their natural appearances, are still almost as widely separated as when we were under the sway of the very defective sexual system. It thus happens that plants of the order *Araceæ* are given various names, according to the different views which botanists take of the value of characters. The botanists of the past age would have called our plant an *Arum*. In the earlier part of the present century, it was regarded as a *Caladium*, and Nuttall refers to it in 1818 as *Caladium sagittifolium*. Rafinesque, about Nuttall's time, placed it in his genus *Peltandra*. Schott, in his revision of *Araceæ* in 1832, created the separate genus *Xanthosma;* and although this is not accepted by some of the best German botanists, who still regard several of Schott's genera as identical with *Caladium*, the division seems to be recognized by American botanists as a sound one, and we have followed their judgment accordingly.

There is a difference among authors as to the orthography. Some have it *Xanthosma*, and others *Xanthosoma*, — Greek words, the first meaning "yellow odor," and the second "yellow body"; but the first is unintelligible, and the application of the last not apparent. Dr. Chapman has *Xanthosoma* in the body of his work, and *Xanthesmia* ("yellow banner") in the index. However, we must leave this question to the linguists to decide, and shall adopt *Xanthosoma* as the name most in favor with our people. *Sagittifolia* is from the resemblance of the leaves to an arrow-head.

The species seems to have no generally recognized common name, but its local name in North Carolina, according to a communication from Dr. Thos. F. Wood, of Wilmington, in that State, is Arrow-leaved Spoonflower. The same English appellation, Spoonflower, was also adopted for the genus by Dr. Curtis, late State botanist of North Carolina, in his "Catalogue of Indigenous Plants."

We have already noted that the plant is used as a vegetable. Dr. Lindley tells us that a starchy substance, called "chou caraib" in the country where it is extracted, is prepared from the roots.

EXPLANATION OF THE PLATE.— 1. Expanded spathe, showing the male flowers in the centre of the spadix. — 2. Scape, with faded spathe. — 3. The same, with portion of the spathe cut away to show the position of the immature fruit on the spadix.

CASSANDRA CALYCULATA.

LEATHER-LEAF; CASSANDRA.

NATURAL ORDER, ERICACEÆ.

CASSANDRA CALYCULATA, Don.—Leaves oblong, mucronate, paler and scurfy beneath, the floral ones oval; flowers in the axils of the upper leaves, small, white; calyx-lobes ovate, acute. Varies with the leaves and calyx-lobes narrower, when it is the *Andromeda angustifolia* of Pursh. (Chapman's *Flora of the Southern United States.* See also Gray's *Manual of the Botany of the Southern United States,* and Wood's *Class-Book of Botany.*)

THE natural order *Ericaceæ*, to which *Cassandra* belongs, is so called from *Erica*, or the well-known Heath of Europe and the Cape of Good Hope. It was for a long time believed that no true Heath was a native of the American continent. A distinguished botanist of the past age, Barton, wrote: "Not a single species of *Erica* is to be met with in this great country; but in place of the 'blooming heather,' Nature has liberally supplied our country with various species of *Andromeda, Vaccinium*, etc., not to mention other genera which are nearly allied to *Erica*." Since Barton's time, however, one true Heath has been discovered in the Northeast in a very few localities; but it is so rare that Barton's remark may be accepted as practically correct. The species now called *Cassandra*, as well as several other genera, were all included in *Andromeda* in Barton's time; but in 1834, this latter genus was rearranged by D. Don. Those, therefore, who wish to examine closely the literature of our plant will have to look for it under the name of *Andromeda calyculata* in all works issued prior to the date just mentioned. Don's divisions are generally accepted now by botanists, although some of them have very few species. In the case of our plant, there is but the single one, although two are generally described in European works; but it will be seen by the description we have adopted at the head of our chapter from Chapman, that American botanists regard the two as one.

Cassandra differs from the true *Andromedas*, particularly in the stigmas and in the anther-cells. These cells are elongated in *Cassandra*, but are short in *Andromeda*, which latter also has a truncated stigma, while the stigma of *Cassandra* is ringlike, with a five-tubercled disk. There are other differences; and a very striking one is the absence of small bractlets under the regular, five-cleft calyx in *Andromeda*, while there are constantly two under the calyx of *Cassandra*.

As the early history of *Cassandra* is connected with *Andromeda*, we may as well stop here to say a few words about the latter name. Andromeda, as Grecian mythology informs us, was the daughter of King Cepheus, of Ethiopia. Being proud of her beauty, she boasted that she was handsomer than even the Nereids, whereat these envious damsels became so enraged that they petitioned Neptune to avenge their wounded feelings. The god accordingly not only devastated the realms of Cepheus by inundations, but also sent a terrible sea-monster, which devoured men and beasts indiscriminately. The oracle of Ammon having announced that these plagues would not cease until the offender had been thrown to the monster, the people compelled their king to chain his daughter to a rock on the seacoast. In this situation Perseus, who had just cut off the head of the Medusa, found Andromeda, and of course delivered and afterwards married her.

The great Swedish naturalist, Linnæus, came across a plant, in the wilds of Lapland, growing under circumstances which suggested this ancient story to his mind, and he accordingly named it *Andromeda polifolia*. Anything connected with Linnæus always pleases those who love wild flowers; and in this anecdote, especially, we seem to be made a sharer of his own thoughts, and are given an insight into his deeper nature which few other anecdotes afford. It shows him as a man of fine, poetic feelings amidst all the details of science, which to some people seem to be intolerably dry, and mere matters of fact. Whenever we look at our pretty *Cassandra*, this incident in the life of Linnæus is recalled to our mind by association with the earlier name of the plant, and we are tempted to invest the incident itself with a personality, and say in the language of Campbell: —

> "I love you for lulling me back into dreams
> Of the blue northern mountains and echoing streams,
> And of birchen glades breathing their balm,
> While the deer was seen glancing in sunshine remote,
> And the deep, mellow crush of the wood-pigeon's note
> Made music that sweetened the calm."

When Don divided the botanical genus *Andromeda*, as before mentioned, he gave to our plant the name of *Cassandra*, still following up the fancy so prettily started by Linnæus. According to Greek mythology, *Cassandra* was the daughter of Priam, the last king of Troy, by Hecuba, one of his wives, — for the old man was a bad polygamist, — and the literal meaning of the name is said to be, "She who inflames with love." The original Cassandra is described as a prophetess, or perhaps a poetess, — little distinction being made between the two in those days, — and her connection with the tragic fate of Agamemnon will be remembered by all. But there appears to be no special reason for giving the name of *Cassandra* to this particular plant, beyond the desire to adhere to the mythological nomenclature suggested by Linnæus.

Our plant is often in flower before the snows have fairly gone. Indeed, it is not difficult for it to do this, as the flower-buds are well advanced before the winter sets in, as shown by our Fig. 1, which was drawn from a specimen gathered in December. A few days of warm sunshine are sufficient to develop the flowers to perfection.

The leaves of *Cassandra calyculata* are very interesting when placed under a lens. The numerous small veins make a sort of net, or rather lace-work, of great beauty, and on these little veins are seen small resinous dots in great numbers, generally three or four times more numerous on the under than on the upper surface. It is not known whether they are of any advantage to the plant as an individual, or whether they are simply of use in that general order of nature which makes all things work together for good. The plant is an evergreen, though with the incoming of winter the lower leaves take on the roseate hue depicted in our plate. As the pretty little waxen-white flowers become perfectly developed, they droop upon their delicately slender stems, and make a pretty wand-like spray, which is really beautiful, and well worthy of study by the devotees of art. In very delicate ornamentation, as in the more precious metals, there are many opportunities for using our plant as a model to great advantage. Even dried specimens, provided they have been dried rapidly, and under great pressure, can be so arranged as to form very pretty wreath-frames for enclosing shells, or similar mementos, or can be made into ornaments of various other kinds.

Dr. Gray gives our plant the common name of "Leather-leaf," but we find no reason anywhere given for this name. Its botanical name, *Cassandra*, ought to be pretty enough to insure general adoption.

The *Cassandra* is a native of Northern Europe and Asia, as well as the United States, but it is remarkable that, while it is rather common from Canada to North Carolina, it is not found west of the Mississippi River. It is extremely common in the barrens of New Jersey, whence we drew our illustration.

EXPLANATION OF THE PLATE.— 1. Branch with buds in December.— 2. Branch in flower in March.

VIOLA SAGITTATA.

ARROW-LEAVED VIOLET.

NATURAL ORDER, VIOLACEÆ.

VIOLA SAGITTATA, Aiton. — Smoothish or hairy; leaves on short and margined, or the later often on long and naked petioles, varying from oblong heart-shaped to halberd-shaped, arrow-shaped, oblong-lanceolate or ovate, denticulate, sometimes cut-toothed near the base; the lateral, or occasionally all the (pretty large purple-blue) petals bearded; spur short and thick; stigma beaked. (Gray's *Manual of the Botany of the Northern United States*. See also Chapman's *Flora of the Southern United States*, and Wood's *Class-Book of Botany*.)

VIOLETS have always been associated with our ideas of early spring. There is scarcely a poet who thinks of spring but refers to the Violets in connection therewith. Says Mrs. Southey: —

"Spring, summer, autumn! Of all three,
   Whose reign is loveliest there?
Oh! is not she who paints the ground,
When its frost fetters are unbound,
   The fairest of the fair?

"I gaze upon her violet beds,
   Laburnums golden-tress'd,
Her flower-spiked almonds; breathe perfume
From lilac and syringa bloom,
   And cry, 'I love spring best!'"

Shakespeare, in making the Duchess of Gloster congratulate her son Aumerle on his being created Duke of Rutland, puts these words into her mouth: —

"Welcome, my son! Who are the violets now
That strew the green lap of the new-come spring?"

And similar allusions to the Violet, as one of the earliest of spring flowers, are very common in the writings of the best authors. Of course these passages refer to the Violet of the Old World, which is not a native of the United States; but most of the poetic associations with the classic Violet are applicable to many of our own species. As regards earliness, our present species may well claim to be admitted as a contestant. The *Viola cucullata*, or Common Blue Violet, has earned the popular name of "Early Blue Violet," but it is questionable whether in a close average, drawn under equal circumstances, the Arrow-leaved Violet would not be awarded the palm. In the vicinity of Philadelphia, it particularly delights to grow on the dry slopes formed by decaying rocks of mica schist, and it is but seldom that those who go out to gather wild flowers after a few warm days at the end of March, or early in April, and who visit these sunny, sheltered spots, return without at least a few Arrow-leaved Violets. Besides being early, it is also continuous. Our drawing was made from a specimen gathered near Philadelphia in May. The flowers often grow larger than those we have chosen for illustration, and in the richness of their blue probably exceed those of all our other species of violets.

One might suppose, from the name "Arrow-leaved Violet," that the leaves would afford a fair, distinctive mark; but these organs often resemble a spoon as much as an arrow-head, and there are, indeed, some other species which have sagittate leaves more frequently than this one. Again, the leaves are often very hairy, and this is especially the case in plants growing on high, dry ground, while in damp situations the leaves are generally quite smooth. Indeed, in most plants the form, or the hairiness of leaves is not relied on now as an exact character in determining species; but these points are useful, in connection with other characters, if we do not forget that they are variable. Although the leaves are not always like an arrow-head, the base is generally abruptly drawn in to form the petiole, — more so than in the other species one is likely to meet with in the early spring season, — and the flower-stems have generally an erect habit, and extend above the leaves, while the Common Blue Violet, with which the Arrow-leaved Violet is most likely to be confounded by beginners in collecting, has the flower-stalks shorter than the leaves.

Though the Violet is essentially a spring flower, —

> "The youth of primy Nature,
> Forward, not permanent; sweet, not lasting!
> The perfume and suppliance of a minute,
> No more,"

as Laertes tells us in "Hamlet," it by no means ceases to bloom in a certain way, but continues to produce seed-vessels during most of the summer season. The flowers which appear in early spring are complete, that is to say, they have not only the organs of generation, but the petals are also perfectly developed, while those which are of later growth produce seeds from apetalous buds, often under ground, and are called "cleistogamic." Although the name for this kind of flower is new, the fact of their existence has been known for many years. Salmon, a writer in the time of Queen Anne, nearly two hundred years ago, says: "The flower of the Violet consists of five leaves, with a short tail. After these, come forth round seed-vessels, standing likewise on their short footstalks, in which is contained round white seed; but these heads rise out from the stalks on which the flowers grew (as is usual in all other plants), but apart by themselves, and being sown, will produce others like unto itself." It is quite probable that it was in the Violets that this strange peculiarity was first noticed; but within the past twenty years, quite a large list of plants with these interesting flowers has been made out. The whole subject has become one of deep study since the writings of Mr. Darwin appeared. It is supposed that the colors of flowers have the purpose of attracting insects, so that pollen may be brought to one plant from another; but the complete flowers of the Violet seem rarely to produce seed. Prof. Goodale says in his "Wild Flowers" that the *Viola sagittata* was never known by him to have seeds from the complete flowers, nor does the writer of this remember to have seen any. The same thing, however, has been believed in relation to some English species; but Mr. Darwin says it is a mistake, as he has seen some fruit in a few cases.

Another very interesting observation has been made in connection with the scattering of the seeds. The capsule is three-valved, and when the seed has matured in all the valves, the latter contract, pressing the grains of seed, which then fly out as a bean flies from between the fingers when pinched. There is a popular prejudice in some parts of England that the Violet "breeds fleas," and this, no doubt, originated from the brown seeds being ejected in the way described. The seeds are about the size of a flea, and any one not looking close enough at the plant to notice the seeds as they are ejected, would be very likely to take the "jumping creature" for a veritable flea.

So far as our observation extends, the Arrow-leaved Violet grows in the Northern States, in open fields or hillsides, in rather dry places; but as we go South, it seems to prefer damp situations. The place of growth seems in some measure to influence its character. Dr. Willis says, in his "Catalogue of New Jersey Plants," that it is generally "slightly pubescent (hairy or downy) when growing in dry soil, and entirely smooth when growing in damp places." Situation and external circumstances often influence form, but frequently there are laws which cause changes quite independent of anything external.

Violets abound in our country, but yet the individual species have a circumscribed limit in many instances. Thus Chapman says of the species to which this article is devoted, that within the area covered by his "Flora of the Southern United States" it is chiefly confined to the upper districts. Its chief territory seems to be Canada and the more northern Atlantic States, and from there west to Michigan, sweeping thence southerly to Arkansas and Florida.

GERARDIA PEDICULARIA.

FERN-LEAVED FALSE FOXGLOVE.

NATURAL ORDER, SCROPHULARIACEÆ.

GERARDIA PEDICULARIA, L.—Smoothish or pubescent, much branched, two to three feet high, very leafy; calyx five-cleft, the lobes often toothed; corolla yellow; the tube elongated, woolly inside, as well as the anthers and filaments; anthers all alike, scarcely included, the cells awn-pointed at the base; leaves ovate-lanceolate, pinnatifid, and the lobes cut and toothed; peduncles longer than the hairy, mostly serrate calyx-lobes. (Gray's *Manual of the Botany of the Northern United States*. See also Chapman's *Flora of the Southern United States*, and Wood's *Class-Book of Botany*.)

THE genus *Gerardia* is so named in honor of one of the most celebrated English botanists, who, as "Gerarde the Herbalist," is constantly referred to in both botanical and horticultural works. To a certain extent, Gerarde may be regarded as the Linnæus of the sixteenth century, and the great Swede recognized the services which his English predecessor had rendered to botany, by dedicating a genus to him when he recast the genera of plants according to the system afterwards known as the Linnæan. What the particular thought was in the mind of Linnæus, which induced him to perpetuate the old English author's name by attaching it to a genus so completely American, does not appear. Modern botanists have made attempts to deprive him of some of his honors; and Rafinesque, whom Dr. Baldwin, in his correspondence with Dr. Darlington, styles a "literary madman," endeavored to make several genera out of *Gerardia*. He calls some of them *Dasanthera*, others *Dasystoma*, others *Eugerardia*, and others, again, *Pagesia*. Some botanists still retain these names. Dr. Gray, however, whom we have credited with our leading description, does not regard the characters which Rafinesque took as generic to be of sufficient importance to divide the genus established by Linnæus, although he retains some as of sectional value. Thus our plant, in Chapman's "Southern Flora" and in Wood's "Class-Book," is classed as *Dasystoma pedicularia*, but in Gray's "Manual" it is *Gerardia pedicularia*, in section *Dasystoma*. *Dasystoma* seems to be from the Greek *dasys*, thick, and *stoma*, a mouth; but unless it be that the corolla is generally of a thicker texture in the species classed as *Dasystoma* than in those placed in the other sections, it is difficult to guess at the application. *Dasystoma* includes all the perennial *Gerardias*.

The *Gerardias* are said to be more or less parasitic on the roots of other plants; but we have been unable to find any clear evidence of the fact in any personal examination, or to find the full proof of it in any published account. All that we have read on the subject seemed to leave room for further observations. One of the reasons given is that no attempts to cultivate it have been successful; but then the same is said of the Trailing Arbutus, — *Epigæa repens*, — which no one pretends is a parasite. Johnson, an English writer, tells us that "*Gerardia pedicularia* was introduced into England in 1826, and all the perennial species can be raised from cuttings as well as by seed, and *G. quercifolia* (a closely related species to *G. pedicularia*) by dividing the rootstocks in spring." If this is the result of actual observation, and not merely assumed because it is the case with perennials in general, it would seem to be established that the *Gerardias* can be grown in gardens. Mr. Thomas Moore, however, of the Chelsea Botanical Garden, London, and one of the best of English practical writers, remarks of the whole family of *Gerardias* that "all attempts to cultivate them in England have failed." But it would be well worth while to try them again. There certainly are large numbers of roots on our plant which have no attachment to other plants, and which must derive nutrition directly from the soil. In the specimen taken up for our illustration, a number of the rootlets had small cellular granules at the end, and it is just possible that these are intended for attachment and suction on roots with which they may come in contact.

An interesting observation in connection with *Gerardia pedicularia*, and the visits of bees to its flowers, has been placed on record, in the "Bulletin of the Torrey Botanical Club" for 1871, by Mr. W. W. Bailey, of Providence, R. I. He found that the humblebees visited the flowers in great numbers; but instead of entering the flowers by their mouths for the nectar, they rested on the upper surface, and then cut a slit, near the base, through which they sipped their sweets. They do this in the petunia, the red clover, the wistaria, and indeed in a large number of other cases in which it is clearly seen that it is difficult for them to enter by the regular "door" of the flower; but as the mouth of the flower of *Gerardia* is so large, it would seem more convenient for the bees to enter by it than to take the trouble of making a slit; and this is the point of Mr. Bailey's observation. But it is evident that it is easier for the bee to stand on the flower and cut it, which it can do without effort, than to sustain itself on the lower part of the mouth and thrust its head down the throat, and it is only reasonable to suppose that insects have the power, to a certain extent, of finding out the easiest ways of doing things.

The flowers have a deliciously sweet odor, which makes them very attractive to the wild-flower gatherer; but they are poor material for bouquets, as they wither very rapidly after gathering. It is remarkable, also, that, in drying specimens for the herbarium, is is almost impossible to preserve the green color. They invariably turn black, even with the greatest care.

> "The foxgloves and the fern,
> How gracefully they grow,
> With grand, old oaks above them
> And wavy grass below!"

These lines of the poet were particularly appropriate to the locality, near Philadelphia, where we gathered the specimen of False Foxglove which served as the original for our illustration. It was in a piece of rather open wood, where the "grand old oaks" of the red, scarlet, and white species waved their branches above, while, somewhat lower down, on ground that was a little more moist, grew the cinnamon fern and numerous sedges, which latter might have been taken for the "wavy grass below." The Fern-leaved False Foxglove generally grows in situations like the one just described in Pennsylvania, in which State it finds itself very much at home, being, perhaps, the most common of the perennial species. Dr. Gray, in his "Manual of Botany," speaks of it as being "common in dry copses." In New Jersey it seems to be found in more open places, and, according to Chapman, it appears to occur in similar locations in dry, sandy soil.

Gerardias in general seem to be confined to the Atlantic States, although some of the annual species are found beyond the Mississippi. Our False Foxglove, however, keeps to the east of this river, where it is found, we believe, in all the States of the Union from Canada to Florida.

The name *pedicularia* was suggested by the great resemblance of the root-leaves to the *Pedicularis*. Our plant has had no common name given to it that we know of, and we have, therefore, ventured to call it the "Fern-leaved False Foxglove."

It blooms in August, and from its branching character keeps a long while in flower.

EXPLANATION OF THE PLATE. — 1. Part of the panicle. — 2. Stamen, showing the awned anther cells. — 3. Seed-vessel and leafy calyx divisions. — 4. Root, with granular-tipped rootlets.

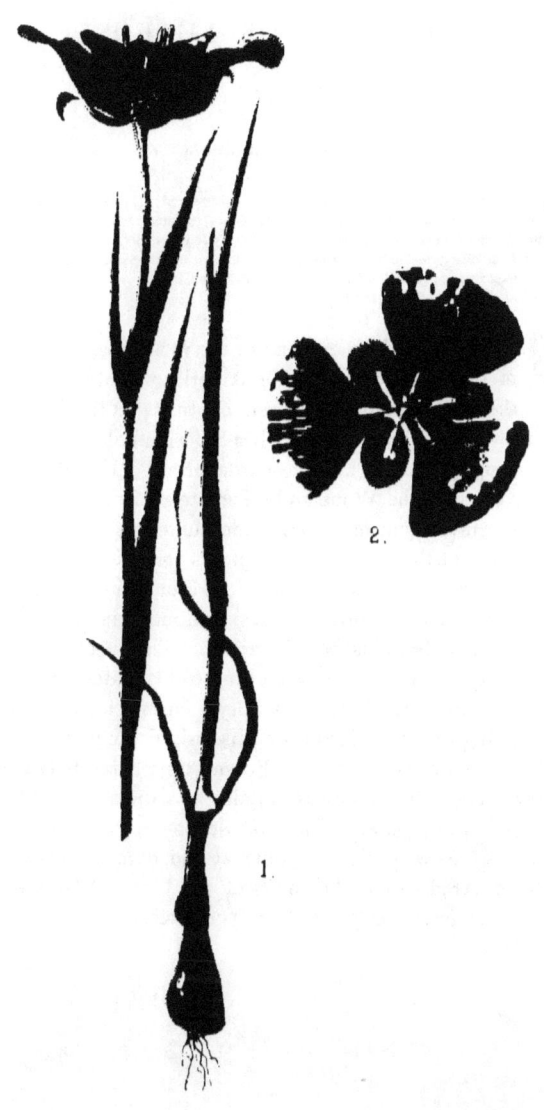

CALOCHORTUS LUTEUS.

YELLOW PRETTY-GRASS.

NATURAL ORDER, LILIACEÆ.

CALOCHORTUS LUTEUS, Douglas.—Stem about three-flowered; leaves convolute-acuminate, shorter than the slender peduncles; sepals oblong, pointed, and recurved at the apex, scarcely shorter than the petals, yellow; petals yellow, broadly cuneate, rounded at the apex, bearded across the base, a roundish, red spot near the middle; anthers as long as the filaments; capsule elliptical; May. (Prof. Wood, in *Proceedings of the Academy of Natural Sciences*, 1868, p. 169.)

WE have in *Calochortus* one of the most interesting genera of plants growing on the American continent. It was only in the beginning of the present century that the first species was discovered by Frederick Pursh, a Siberian botanist, who came to Philadelphia in 1799, and was gardener to W. Hamilton, whose grounds are now the Woodland's Cemetery of that city. Pursh was a very intelligent man, and made numerous excursions into various parts of the country. The plants collected by him during these excursions afterwards formed the foundation of his "Flora of North America," a work published in London in 1814, in which the genus *Calochortus* is first described. The species he discovered, *C. elegans*, was found, according to his statement, in what was then the great Louisiana Territory. No species has ever been seen growing wild this side of the Mississippi, the numerous ones that have been discovered since Pursh's time being native to the country between the Rocky Mountains and the shores of the Pacific. A few are found in Mexico, and others extend northward to Oregon. New species continue, at the date of this writing, to be discovered within the limits of the United States, so that the exact range of the genus is not yet determined.

Although *Calochortus* is exclusively American, it is yet not distantly connected with the tulip of the Old World, and is also closely related to the *Erythronium, Fritillaria,* and some other American genera, with which it unites in giving interest to the great tribe of *Tulipeæ*. One striking difference from any of its allies, however, will be noted by the most casual observer. The flower-cup of the common tulip seems to be formed of six petals; but in reality, three of these apparent petals are sepals, for the flower is formed on a ternary plan. The sepals are, however, so petal-like that there seems to be no calyx in the ordinary sense of the term. In *Calochortus*, however, there is seen to be an approach to the general condition of a complete flower. The three outer leaves or sepals, although still colored somewhat as the petals are, as we see in our full-face view, Fig. 2, are so much smaller than the inner ones forming the corolla that no one would have any difficulty in at once deciding the distinctness of the two series of floral envelopes. This observation is particularly worthy of the attention of students interested in a comparison of structure; for with this separation of the calycine from the corolline system, we see that *Calochortus* approaches another order, of which our *Tradescantia*, or Spiderwort, is a familiar example. The glaucous, sub-fleshy leaves of most of the species of the two families also somewhat resemble each other in character, and the tendency to the production of silky hairs in the stamens of *Tradescantia* finds some parallel in the tufty hairs often produced on the petals of *Calochortus*. These characters are, however, mere appearances, and would not weigh much in regular systematic botany; but they will be of some value to our readers, many of whom are interested in the general tendency of relationship, as well as in the more exact studies.

The name *Calochortus* is from the Greek *kalos*, pretty, and *chortus*, grass. The leaves of most of the species have a grassy appearance; and in view of the beautiful flowers on so grass-like a plant, when the real grasses have no such beauty, the name is a very appropriate one. It is to be regretted that the translation of this name, "Pretty-Grass," did not become part of the language of the people; but "Butterfly Weed," "Mariposa Lily," and "Wild Tulip" have become so common in California, that there seems to be hardly any chance for the plant ever to get a distinctive appellation of its own. In the hope, however, that Pretty-Grass may yet become popular, we shall use this name in our present chapter.

Calochortus luteus, the Yellow Pretty-Grass, was first discovered by Mr. Douglas, who, in 1831, collected in California for the Royal Horticultural Society; and in the society's garden at Chiswick, near London, it flowered about the year named.

As Professor Wood says, in the description we have quoted, it has generally three flowers on a stalk; but as it sometimes comes with but a single flower, it will serve a good purpose to illustrate a plant in that condition, as showing the range of variation. It will also be seen, by reference to the plate, that our plant varies in another particular from Prof. Wood's description, the red spot being, not in the middle, but rather lower down on the petals. Our drawing is from a cultivated specimen, brought from California by Mr. Edwin Lonsdale, of Germantown.

It has been a matter of surprise that so pretty a flower, introduced to Europe under the auspices of its leading horticultural society, should be so rarely met with in cultivation. But this is chiefly owing, perhaps, to the necessity of importing roots direct; for according to our experience in raising the plants from seed, it must often take many years to procure flowering bulbs in that way. Seeds that we have sown only made bulbs the size of grains of wheat the first season; and though these bulbs produced leaves annually, they had not much increased in size after several years. We have heard of one grower, who soon had flowering bulbs from seed, but we think this success must be rare. Another difficulty is this, that the roots do not seem to increase rapidly by offsets, as some bulbs do. The plant from which we took our drawing made but two small bulbs, much smaller than the original Californian root, and these came out at the points represented as two small swellings in the plate, from which it will be seen that they were on the stem rather than on a part of the old bulb.

Almost every traveller who goes through California in the late fall of the year writes to Eastern friends of the great beauty of the plains and foot-hills when glowing with the gold of the Mariposa Lilies, which we take to be the species we now illustrate. The phenomenon is especially noticed by those who go through the Sacramento Valley, where, to judge from all the accounts given, it seems to find itself the most completely at home. In cultivation it would probably not be early enough for our outdoor gardening; but it will be an excellent thing for pot-culture in windows or green-houses. In this respect it has one very great advantage. We have spoken of the connecting link between it and the *Tradescantia*, or Spiderwort; but it will not do to compare the endurance of the petals in the two flowers; for while the Spiderwort lasts only a few hours, the Yellow Pretty-Grass will endure for a long time. The flowers on the plant from which we took our drawing kept open a week, and other growers have even had a still more favorable experience. The editor of the " London Garden," July 1, 1876, says: " We have so long considered the Mariposa Lilies somewhat delicate and fragile, owing to seeing them till recently represented by very poor specimens, that we are agreeably surprised at finding they keep for a considerable time in water, and open their large, gay, yet delicately marked blooms freely. The ones before us are of a fine dazzling yellow color, like *Calochortus venustus*, but of the most dazzling yellow, with brownish-crimson pencillings and markings." We quote this because it evidently refers to the species we have now before us.

EXPLANATION OF THE PLATE.— 1. Bulb with complete plant and side view of flower.— 2. Full-face view of flower.

IRIS VERSICOLOR.

BLUE FLAG.

NATURAL ORDER, IRIDACEÆ.

IRIS VERSICOLOR, Linnæus. — Stem stout, angled on one side ; leaves sword-shaped (three quarters of an inch wide); ovary obtusely triangular with the sides flat ; flowers (two and one half to three inches long) short peduncled, the funnel-form tube shorter than the ovary; pod oblong, turgid, with rounded angles. (Gray's *Manual of the Botany of the Northern United States.* See also Darlington's *Flora Cestrica*, Wood's *Class-Book of Botany*, and Chapman's *Flora of the Southern United States.*)

THE genus to which the Blue Flag belongs was called *Iris*, which is the Greek name for rainbow, on account of the brilliant hues displayed by the flowers of some of the species. This brilliancy of color is characteristic of all the American species comprised in the genus, and the plant which we are about to examine does special honor to the name. The hues of the flowers of the Blue Flag are not, indeed, exactly those of the rainbow, but they are quite as varied ; and in this respect the specific name of the plant, *versicolor*, is very appropriate. The great beauty of the *Iris versicolor* has always won admiration, and has frequently called forth happy lines from the poets. Longfellow, with the popular idea of the relationship of our plant to the Lily present in his mind, thus sings of it : —

" Beautiful Lily, dwelling by still rivers,
 Or solitary mere,
Or where the sluggish meadow brook delivers
 Its waters to the weir,

Thou laughest at the mill, the whir and worry
 Of spindle and of loom,
And the great wheel that toils amid the hurry
 And rushing of the flume.

> Born to the purple, born to joy and pleasance,
>   Thou dost not toil nor spin,
> But makest glad and radiant with thy presence
>   The meadow and the lin.
>
> The wind blows and uplifts thy drooping banner,
>   And round thee throng and run
> The rushes, the green yeomen of thy manor,
>   The outlaws of the sun.
>
> The burnished dragon-fly is thy attendant,
>   And tilts against the field,
> And down the listed sunbeam rides resplendent
>   With steel-blue mail and shield.
>
> Thou art the Iris, fair among the fairest,
>   Who, armed with golden rod,
> And wingèd with celestial azure, bearest
>   The message of some god.
>
> Thou art the Muse, who, far from crowded cities,
>   Hauntest the sylvan streams,
> Playing on pipes of reed the artless ditties
>   That come to us in dreams."

The Blue Flag is, indeed, one of the most beautiful of all swamp-loving plants; and a large tract covered with it, while its flowers are in full bloom, as often seen in May or June, is one of the most pleasing sights in nature.

The evident relationship in the poet's mind between our plant and the Lily, and his allusion to the wind which uplifts its "drooping banner," naturally lead us to a consideration of the structure of the flower. In this respect the reflexed sepals, or leaves of the outer division of the perianth, first claim the attention of the student, as they are characteristic of many of the species included in the genus. These sepals turn outward and downward, while in the neighboring order of *Amaryllidaceæ* the floral parts which answer to them have rather an inward direction. From the true Lilies the *Iridaceæ* are widely separated in the natural classification, although the first cause of this wide differentiation is apparently of no great moment. For this very reason the study of the structure of the flower is all the more interesting, as it serves to show on what seemingly small changes hinge the most wonderful divisions of the vegetable kingdom.

It is well to keep a Lily flower in view while studying the manner in which the flower of an Iris is built. In the true Lily the perianth is free from the ovary, which latter is therefore called superior in botanical language; while in the plants of the order *Iridaceæ* the perianth is united to the ovary, which is therefore inferior. (See Fig. 3.) As the *Iridaceæ* as well as the Lilies are endogens, they have their parts in threes in their normal condition. Thus there are three sepals and three petals in the Lily; but in the rhythmical development of growth the two verticils have apparently been arrested together, and both sets are therefore so much alike that there seems to be no distinction between them. It is impossible to tell the sepals from the petals, and it would be quite as correct to say that the perianth of the Lily is composed of six petals, as to say that it is composed of six sepals. In the Iris the perianth also consists of twice three parts, but it is evident that the verticils have been influenced separately. The three leaves which form the lower verticil, and which may be called sepals, although they are purely petaloid, have broad blades and turn downwards; while the second verticil has assumed the shape of comparatively small petals which incline upwards. In the stamens we note a still more remarkable difference between the Lily and the Iris than in the perianth. The Lily has six stamens, and these, like the leaves of the flower-cup, are formed of two verticils of three each. It is difficult, however, to distinguish the two series from one another, as they have both been caught, in very close succession, by the same growth-wave; but if we watch the development of these six stamens, we find that three of the anthers expel their pollen somewhat before the other three, and from this fact we learn that they really represent two stages of growth. The pistil, in like manner, was originally in a ternary condition, but the normally distinct parts have been so united that their trifid character is only revealed by the three-parted stigma at the apex. If now we revert again to the Iris, we shall find but a single verticil of stamens,— three only; but these are in their proper situation, or in other words, they alternate with the petals, and bend back over the median line of the sepals. The other three which we might expect to find, judging from the analogous structure of the Lily, have wholly disappeared. As regards the pistils, they would almost seem to be wanting, at first sight, and in the place which they ought to occupy, we notice a petaloid, three-parted structure in the centre of the flower, inside of the stamens. On closer examination, however, we discover that these structures are really the pistils, and that they have coalesced with another set of three bodies which might have formed a second verticil of three stamens, but which are still petaloid in character. From the morphological lessons we have already learned, we can now understand what has become of the second verticil of stamens. These organs have evidently been united with the next verticil, or the pistils, and thus we have the beautiful petaloid pistils, which give such a peculiar character to the Iris.

Our plant abounds in Maine in the East, and in Minnesota in the West, and is as much at home in Arkansas as in Florida, and throughout the whole of the vast territory of which these points indicate the limits.

According to Lindley, the Blue Flag is a "diuretic, purgative, and emetic." Bartram, in his "Travels," tells us that it was in great favor with the Indians as a powerful cathartic; and indeed he intimates that its wide distribution is in a great measure due to the estimation in which it was held by them for medicinal purposes. It has also been found useful in cases of dropsy. In overdoses, it causes nausea, similar to sea-sickness.

EXPLANATION OF THE PLATE.— 1. Flowering stem, proceeding from the terminal bud of a rhizome of last year's growth. — 2. Branchlet of the flower stem, with expanded flower. — 3. Faded flower.— 4. Cross section of the ovary.

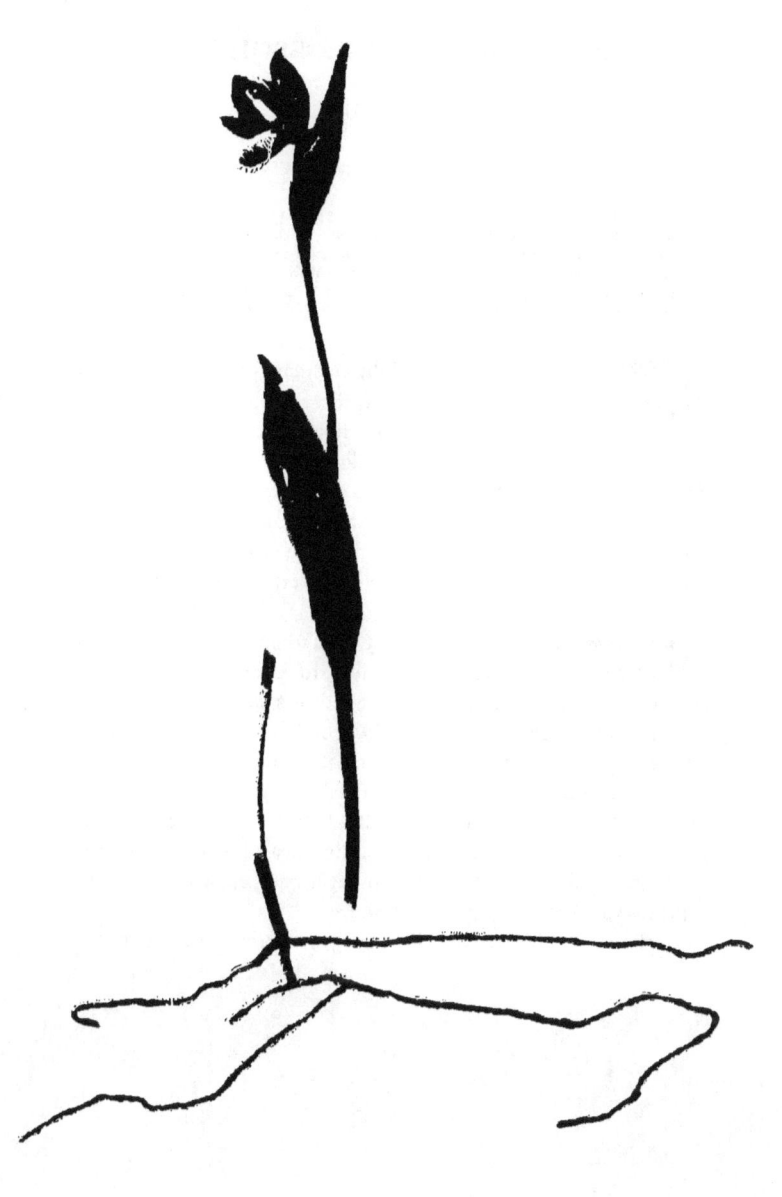

POGONIA OPHIOGLOSSOIDES.

SNAKE-MOUTH.

NATURAL ORDER, ORCHIDACEÆ.

POGONIA OPHIOGLOSSOIDES, Nutt. — Root of thick fibres; stem (six to nine inches high) bearing a single oval or lance-oblong leaf near the middle, and a smaller one or bract near the terminal flower, rarely one or two others with a flower in their axil; lip spatulate below, appressed to the column, beard crested and fringed; flower, one inch long, sweet-scented. (Gray's *Manual of the Botany of the Northern United States*. See also Chapman's *Flora of the Southern United States* and Wood's *Class-Book of Botany*.)

ORCHIDS seem at first sight calculated to shake our confidence in the reliability of the morphological doctrine, according to which all the parts of a flower are but modifications of simple leaves. On closer investigation it will be seen, however, that hardly a better illustration of the truth of this doctrine could be found than is offered by a comparison of our present species, *Pogonia ophioglossoides*, with some other species of the same genus, and more especially with *P. verticillata* and *P. pendula*.

In *P. pendula* the stem is leafy, and there are a number of axillary flowers (one to four, according to Gray; three to seven, according to Chapman), but these flowers are far from being showy. The stem of *P. verticillata*, on the contrary, is naked (excepting some small scales at the base), and there is only one whorl of leaves at the summit, at the base of the reddish-brown flower. In *P. ophioglossoides*, finally, there is one leaf acting as a sheathing scale at the base, another near the middle of the stem, and again a smaller leaf or bract higher up, and above this a pretty rose-colored terminal flower.

In the case of *P. pendula* it might therefore be said that the

Great Architect had not got far beyond the foundations of his work in making a *Pogonia*. The vegetative force seems feeble, and spends itself in often-repeated attempts; hence, small leaves and insignificant flowers are scattered all along the stem. But in *P. verticillata* the force exercised is evidently greater, not only in amount, but also in degree, and its action is more concentrated. The stem, therefore, instead of slowly elongating, and sending out a leaf and a flower here and there, rapidly draws in its spiral coils, thus producing only a verticil or whorl of leaves, and annihilating all tendency to flower in the axils; after which it makes another growth, and then another sudden arrest and coil, resulting in a large, single flower. Coming now to *P. ophioglossoides*, we find that the acting force was intermediate in intensity. Having coiled up the primordial leaves to form the flower-stem, the force was not powerful enough to arrest the formation of the leaves suddenly, and it therefore still left them somewhat scattered. Of the three leaves thus produced the lowermost is little more than a sheathing scale. The next or largest one shows by the groove down the stem opposite, as seen in our illustration, how very near it came to diverging still more than it actually does from the interior leaves, out of which the stem is formed; and the upper one, by its greatly reduced size, reveals the fact that the force employed in arresting the elongating growth, and in working up all the separate parts into a flower, is now in active operation. Thus we see how an exceedingly beautiful structure is built up from a few rough and simple materials.

In distinguishing the genera of Orchids, the relative differences in the sizes and forms of sepals and petals are taken into consideration, as well as the relative forms of the petals themselves. The lip is often very characteristic, and almost alone will enable the botanist to build a genus on it. Our present plant was regarded by Linnæus as an *Arethusa*, and as such it is described in all old works; but in this last-named genus, the sepals and petals are united at the base, while in *Pogonia* they are all distinct. There are other differences, but this one will strike the most cursory observer, and is well fitted to illustrate the point we wish to make, — that great apparent differences are often the result of very slight causes.

In our plant the lip is prettily bearded, and this suggested the name *Pogonia*, *pogon* being Greek for "beard." The specific name, *ophioglossoides*, is derived from the resemblance of the leaves to the fronds of an *Ophioglossum*, a cryptogamic genus allied to the ferns. The English name "Snake-mouth" seems to be adopted by a great many writers, although we never heard our plant popularly thus called. One might suppose that there was some resemblance to a snake's mouth in the flower, but there is none, and the name is evidently suggested by the relation which the specific appellation bears to a snake, *Ophioglossum*, from which it is derived, meaning "serpent's tongue."

Orchids are singularly circumscribed in their geographical ranges; but our present species, where it exists at all, is usually found in great abundance. It grows generally in bogs, among sphagnum and sedges, and in places so wet that those who go out collecting in patent-leather shoes have generally to be satisfied with admiring from a distance. Sometimes a bog will be perfectly ablaze with the bright purple blossoms, and we have frequently seen this beautiful sight, especially in the State of New Jersey. Our specimen was of Massachusetts growth, and rather smaller than is usual in more southern locations. The thready roots creep freely through the decaying moss and mud among which the plant grows, and are so small that those who collect for cultivation experience great difficulty in finding them. For this reason, it is necessary to take them up as they are going out of flower.

The interest in Orchids has of late years been particularly deep, on account of Mr. Darwin's papers on Orchid fertilization. The flowers of the Orchids are generally so constructed as to be unable to fertilize themselves, and they seem to be in a great measure dependent on insects. Mr. Darwin, speaking of our present species in this connection, says: "The flowers of *Pogonia ophioglossoides*, as described by Mr. Scudder, resemble those of *Cephalanthera* (a Mediterranean species) in not having a rostellum (that is to say, a beak), and in the pollen masses not being furnished with caudicles. The pollen consists of powdery masses, not united by threads. Self-fertilization seems to be effectually prevented, and the flowers on distinct plants must intercross, for each plant bears generally but a single flower." It will be observed, however, that Mr. Darwin argues only from the facts to be derived from a study of the *structure* of the flower, so that there is yet room for the student to make original observations, based upon its *actual behavior*. It is worthy of note that, with all the supposed advantages of cross-fertilization, there are not many families of Orchids in this country, nor indeed are the plants spread over wide districts. Of the genus *Pogonia*, there are not many species, and its only close allies in America are *Calopogon* and *Arethusa*, of which there are fewer species than of *Pogonia* itself. If these flowers are so beautifully colored for the especial purpose of attracting insects to their charms, they seem to profit so little by the arrangement that one might be pardoned for suggesting they would have been better off in an humbler garment. The lines of Paulding seem very applicable to them: —

> "Be thine to live and never know
> Sweet sympathy in joy or woe;
> To see Time rob thee, one by one,
> Of every charm thou e'er hast known;
> To see the moth that round thee came
> Flit to some newer, brighter flame,
> And never know thy destined fate
> Till to retrieve it is too late!"

The Snake-mouth is found from Canada West to Wisconsin, and southward to Florida. It flowers in June. There is nothing recorded of its value in the arts.

CLEOME PUNGENS.

PRICKLY CLEOME, OR SPIDERFLOWER.

NATURAL ORDER, CAPPARIDACEÆ.

CLEOME PUNGENS, Willdenow.—Clammy-pubescent; leaves five to seven foliate, long-petioled; leaflets lanceolate, acute, serrulate; lower bracts trifoliolate, the upper ones simple, cordate, ovate; stipules spiny; capsule smooth, shorter than the elongated stipe; seeds rugose; stem two to four feet high; petioles more or less spiny; flowers showy, purple, changing to white. (Chapman's *Flora of the Southern United States.* See also Wood's *Class-Book of Botany.*)

THE Prickly Cleome, a beautiful wild flower from the southern shores of the United States, is an object of curiosity, even to the ordinary observer, from the resemblance which the flowers bear to an insect with erect wings and long legs and tentacles. The resemblance, indeed, is not so striking as in the flowers of some orchids, but sufficiently so to produce an odd effect. To more scientific observers, however, and especially to those who like to examine structure closely, by comparing allied plants with one another, this species offers other points of peculiar interest. If the four petals and four sepals were not all turned in one direction, and if it were not for the general appearance of the seed-vessel, one would suppose at first sight that the plant belonged to the *Cruciferæ*, or cabbage tribe; but in that family four of the six stamens are invariably long, and the remaining two invariably short, while in *Cleome pungens* the whole six are of equal length. The most striking difference, however, will be found in the ovarium, ultimately the seed-vessel, which in *Cleome pungens* is borne on the end of a very long stalk. These, then, together with some other more minute but essential peculiarities, will show at once the difference between our plant and the *Cruciferæ*.

But there are still other plants belonging to the same natural order with *Cleome* which have the petals ranged with more regularity around the axis, in which the ovary is borne on a much shorter pedicel, and which, therefore, point to a much closer relationship to cruciferous plants than that suggested by the plant to which this chapter is devoted. Nor is this suggestion deceptive; for the order *Capparidaceæ*, to which our species belongs, is closely allied to *Cruciferæ*, and quite as much so, if not still more closely, to *Resedaceæ*, or mignonettes. With the violets it also has some affinity, and it has therefore been classed near them. The *Cruciferæ* comprise a very large number of genera and species, while the *Capparidaceæ* and the *Resedaceæ* have each but a very few; and it will be well to look for "missing links" in their development, as it is not improbable that both of them had cruciferous parentage.

A fair key to the structure of our species is supplied by the numerous bracts among the flowers. In the true *Cruciferæ* there is nothing but a naked flower-stalk bearing pedicels and flowers, and the bracts are entirely wanting. We see, also, that the foliaceous system of our plant is very well developed, and this, in plants not absolutely acaulescent, generally implies a corresponding activity in the axis or stem. In other genera of *Capparids*, where there are no bracts, there is scarcely any pedicel to the flower, or to the ovary, and the resemblance to true *Cruciferæ* is every way closer. In this species, however, the tendency to produce stems is so strong that even the petals are stalked, while the stamens have long, drawn-out filaments, and the same force has projected the ovary far beyond the point usual in flowers.

We also see the operation of rhythmical growth, or of the law of acceleration and retardation, as it is sometimes called, in producing certain other results. The pedicel is really a branch, which has started to grow from the axial bud at the base of the leaf. Its excessive slenderness, as compared with the main stem, shows us at once that its vegetative growth has been severely checked, although we notice at the same time that its power of elongation has not been interfered with to the same degree as its power for increase in thickness; but all at once it receives a sudden check to form the calyx, in which latter there is no sign of any elongating or axial growth. The growth-wave is then again somewhat accelerated in a forward direction, and produces the pedicellate or clawed petals; and finally, it is once more accelerated to a still greater degree for the production of the stamens. Thus we see that in the varying degrees of intensity in the growth-wave, and in the degree of rapidity with which the spiral line, along which the vegetative force acts, is drawn in or coiled up, we have the clew to this singular structure, and in some respects the measure of the difference between it and its allies. This, indeed, is true of all plants, but in few is it so well illustrated as in the *Cleome pungens*.

The great beauty of our plant makes it a desirable one to cultivate. It thrives well during the summer in any ordinary garden ground, and indeed the hotter the weather, the better it thrives. It grows about four feet high, and as it branches freely from the sides of the main stem, it makes a showy and symmetrical bush. In ordinary wild locations, unless it happens to find itself in extra rich ground, it does not usually grow more than two feet high. It is an annual in cultivation, although classed as a biennial in most descriptions.

Prof. Grisebach, a noted botanist, does not believe that plants were created all in one place, and that they have wandered over the world from one home. He is of opinion that there have been many centres of creation. But whether this be so or not as to the first appearance of plants on the earth's surface, it certainly seems to be true that our modern races have home-centres, and that from these they have wandered, and still continue to wander, farther and farther away. Most of the species of *Capparidaceæ* are tropical or semi-tropical, but they are continually extending their boundaries. Our *Cleome* is believed to have crossed over to Florida from the West India Islands, and it is probable that it has been introduced into the State named only within recent times, as it is not mentioned in Torrey and Gray's "Flora of North America," which was published in 1840. Prof. Wood gives, in a general way, "the South" as its location, while Chapman places it in "Florida and westward." Mr. George D. Butler, in a note to the "Botanical Bulletin" (now the "Botanical Gazette"), reports it as having already crossed the Mississippi to Arkansas, and there is no doubt but future generations will find it completely across the continent. According to Mr. Martindale, it occasionally appears along the shores of the Delaware, being, no doubt, brought there in the ballast of vessels.

The name of the genus, *Cleome*, is said to be derived from the Greek verb *kleio*, to shut, in allusion to the fact that the style and the filaments, which lengthen faster than the petals, burst through the latter while they are still closed, and while the stigma and the anthers are still enfolded by them. Don says that the name "was first used by Theodosius," and from him adopted by Linnæus. A species is often found described as *C. spinosa*, but this is now thought to be identical with *C. pungens*, which has the right of priority, as far as the name is concerned. Our species has no English name, but a translation of its botanical appellation, "Prickly Cleome" (pronounce clay-om-ay), will, no doubt, be acceptable, unless, indeed, "Spider-flower," which we have heard suggested, should be adopted in preference. We cannot, however, endorse this name, as it is so like Spiderwort, which has already been appropriated by *Tradescantia*.

From a utilitarian point of view, the Prickly Cleome is useless, but it is to be hoped that its beauty will be considered a sufficient reason for its existence.

2.

ACTINOMERIS SQUARROSA.

SQUARROSE ACTINOMERIS.

NATURAL ORDER, ASTERACEÆ (COMPOSITÆ).

ACTINOMERIS SQUARROSA, Nuttall. — Stem somewhat hairy and winged above, four to eight feet high; leaves alternate or the lower opposite, oblong or ovate-lanceolate, pointed at both ends; heads in an open corymbed panicle; scales of the involucre in two rows, the outer linear-spatulate reflexed; rays four to ten, irregular; achenia broadly winged; receptacle globular. (Gray's *Manual of the Botany of the Northern United States*. See also Chapman's *Flora of the Southern United States* and Wood's *Class-Book of Botany*.)

THE species which we now illustrate is not one that will attract by its beauty, if by beauty we understand mere color. But to the true lover of nature, or to the botanical student, it will be acceptable, for there are few which are so instructive, or which afford so many lessons. The plants called *Umbelliferæ*, such as the carrot, parsnip, celery, and so on, are nearly related to the *Compositæ*, of which our plant is a representative. Yet we must look at these two orders in the light of morphological law to see the relationship; for in general appearance they are so different that it has been found necessary to place them somewhat widely apart in the systematic classification of the orders. When we examine a plant of the umbelliferous order, we see that the flower is composed of five distinct petals, and of five stamens, each of which is likewise separate and distinct from the other; but in the flowers of the *Compositæ* the normal five-petalled corolla is united into a tubular one, and the anthers are also united together by their edges, so that the pistil, as it grows, has to push through the united mass. Now morphology teaches us that all the parts of a plant are normally leaf-blades, and that from the various degrees of union or of separation, the degree of individualization or consolidation of these original parts, result the different characters which are exhibited by the different parts of a plant. And we can see by studying such plants as the one we are describing that not only is this true of individual plants, but that differences between species, genera, and orders depend on the same laws of individualization and cohesion, or on the varying degrees of rapidity with which development takes place. An umbelliferous plant is simply a Composite, with less tendency to an arrestation of its axial growth, and a consequent union of parts. The seeds of Composites often have so great a resemblance to the seeds of umbelliferous plants that it is difficult to distinguish the order by them alone. The seeds or "achenes" of the present species greatly resemble those of the parsnip, and of similar umbellifers, in the broad marginal wing on the edges, as seen in the half-mature achene in Fig. 2, and its cross section Fig. 3, and this resemblance is peculiarly conspicuous when the seed is ripe.

In old times *Actinomeris* was thought to belong to *Coreopsis*, and as a member of this genus the first species known to Europeans is therefore described by Willdenow. Nuttall, however, showed that it is much more nearly related to the *Helianthus* or Sunflower, although there are many points of difference between the two, the one which will strike the most casual observer being the small number of the ray-petals, as already noted. The principal flower on our plate is represented with eleven rays, but this is unusual; six, and often only four, being found much more frequently. The name of the genus, *Actinomeris*, is based on this fact, *aktin* being Greek for "ray," and *meris* for "part," the compound thus signifying that the flowers are only "partly rayed."

We may sometimes notice a regular current of air moving in the atmosphere with scarcely any apparent vibration of its wave, while at other times the current is extremely fitful, — now calm and flowing in one direction, now violent and coming in gusts from "all ways at once." The same varying waves can be noticed in the growth-currents of plants, and in this species we have an illustration of the fitful current. We see that the growth-force still retained considerable power in the first effort at forming a flower-bud in the axil of the lowest leaf, and that but little of this power was diverted to advance the reproductive development. The next bud in the series started with a good amount of growth-force, but was suddenly arrested, and the growth-force being converted into reproductive force at this point, the result was that the flowers here formed were stronger, and therefore opened sooner than those in the axil below, which had been produced by a weaker developing power. The next wave after this vigorous arrest moved slowly, and resulted, at the next bract, in a very weak head of flowers; but before its final arrest the growth-force again gathered more strength, and a much stronger cluster was therefore the last achievement of its activity. The student can thus trace the fitfulness of the growth-wave through the whole development of the plant. We see it distinctly in the leaves, which sometimes appear in threes, sometimes opposite, and sometimes alternate, all on the same plant. Our Fig. 4 is a part of the stem with an opposite pair of leaves, while the bracts on the flower-branch, Fig. 1, are alternate. The leaves run down below the point of junction with the stem, or, as the botanists say, they are decurrent, and this gives the stem a four-angled appearance, with green, leafy wings on the angles.

There are several species of *Actinomeris*. The present one, *A. squarrosa*, has been long known, and is described by Linnæus as *Coreopsis alternifolia*. As a cultivated plant, it has been in English gardens for perhaps two hundred and fifty years, and it must have been among the first of our native flowers to make the acquaintance of the botanists of the Old World.

In *Actinomeris squarrosa* the specific or last name signifies jagged or spreading, in reference to the spreading tips of the involucral scales. Not having attracted much popular attention, it seems to have gained no common name.

The geography of this species is of peculiar interest. It seems to be confined to an inland strip of country, but why it has not extended farther north and east is a problem. Most of our botanists give Western New York as its eastern boundary. It was included in Torrey's catalogue of the plants of New Jersey, but this was supposed to be an error. In recent years, however, it has certainly been found at Paterson and at Mont Clair in that State, according to Willis and the " Bulletin of the Torrey Botanical Club." It is found in Pennsylvania, occasionally up to the Delaware River, near Philadelphia, but has not crossed. Chapman says it grows in Florida and northward to North Carolina; but its great home-centre seems to be in Ohio, Michigan, and the adjoining Southwestern States. Thence it is more sparingly found, until it loses itself in the deserts of Western Kansas and Nebraska. As other species are found in the Southwest, we shall probably have to look in that direction for its genetic home.

Our plant commences to bloom rather early for an autumnal flower, but its blossoms are continued far into the fall of the year. As we have before said, it was introduced many years ago into English gardens, although it seems to be rare there now, and we know of no attempt to cultivate it in our own country. It is by no means a showy plant, but still it deserves a place in the flower borders of the real lover of nature, on account of the many valuable lessons it teaches, some of which we have briefly alluded to. It seems to be a great favorite with certain coleopterous insects, which seek out and greedily devour the flowers, although there may be an abundance of others to feed on.

EXPLANATION OF THE PLATE.— 1. Part of a flower-stalk.— 2. Achene, half mature, with two divergent calyx-teeth.— 3. Cross section of the same.— 4. Portion of flower-stalk from about midway, showing a pair of opposite leaves, which it sometimes produces.

CLAYTONIA VIRGINICA.

SPRING-BEAUTY, NOTCH-PETALLED CLAYTONIA.

NATURAL ORDER, PORTULACACEÆ.

CLAYTONIA VIRGINICA, L.—Root a deep tuber; stems six to ten inches long, simple; leaves mostly two, linear-lanceolate, an opposite pair near the middle of the stem, from three to nine inches in length; flowers pale red, with purple veins, usually six to twelve, or even fifteen, in a loose, simple, terminal raceme. (Darlington's *Flora Cestrica*. See also Gray's *Manual of the Botany of the Northern United States*, Chapman's *Flora of the Southern United States*, and Wood's *Class-Book of Botany*.)

IN the early part of the last century, when Linnæus had just succeeded in reducing botany from a mass of confusion to something like order, the native flowers of our own country were beginning to attract the attention of the scientific men of Europe. The Dutch botanists had established close relationship with Americans, and as early as 1739 Gronovius published at Leyden a "Flora Virginica," the figures and descriptions for which were furnished by John Clayton, of Virginia, who did wonders, for that early period, in making our native plants known. At the same time, John Bartram, farmer, physician, mechanic, and botanist, who lived in Pennsylvania, was in active correspondence with England, and sent roots and seeds to his friends there.

In view of the eminent services which Clayton rendered to American botany, it is very fitting that a genus so interesting and so peculiarly American as the one to which our plant belongs should have been named in his honor. Nor is there much danger that the monument thus erected to Clayton's memory will ever be destroyed, as has been the case with so many similar monuments dedicated to other botanists, for the genus *Claytonia* is so distinct in character that there is little chance of its ever being merged in some other genus.

There are about twenty species of *Claytonia* known at present, according to the most recent enumeration, and these are chiefly natives of Northwestern America, or of Siberia. The whole order of *Portulacaceæ*, indeed, to which our genus belongs, has but few representatives in the European flora. At the time of Linnæus only two species of *Claytonia* were known, — *C. Sibirica*, from Eastern Asia, and our American plant, the first acquaintance with which must have been a delightful surprise to Europeans. Certain it is that they took great pleasure in it after it had once been made known to them. Dr. Fothergill, in a letter written to John Bartram in 1772, boasted that he possessed what he believed to be the only plant of *Claytonia Virginica* in "all England." Old Peter Collinson was before him, however, for in Darlington's "Memorials" we find a letter, dated April 10, 1767, in which he records the fact that his *Claytonia* had flowered on April 5.

Many tuberous-rooted plants produce new tubers every year, and the old ones die; in others, on the contrary, the tubers, as thickened root-stocks, live on from year to year, and continue to increase in size. We have had no opportunity, however, to ascertain the facts in the case of our plant, but as far as we have examined the roots, they seem to us to have very much the appearance of being perennial. They are usually very deep in the ground, and we dare say that, to many hundreds of those who go out to gather and admire wild flowers, we shall here be introducing the roots for the first time.

In English works we find the plant spoken of as the "Notch-petal'd Claytonia"; but in American works it is called "Spring-Beauty," with a unanimity quite unusual in the application of common names. It is certainly worthy of the name, for it is very beautiful, and although not the first to flower, it is yet among the earliest harbingers of spring, and gives a great charm to woods and shaded places in April and May. We gathered the specimen from which our drawing was made in the early part of May in a wood, where it grew in company with anemones, ranunculus, and other early plants.

The Spring-Beauty has not as yet succeeded in attracting the attention of philosophers, physicians, or economists. Its next of kin, however, the Common Purslane, is sometimes boiled and eaten; and it is said that the leaves of our plant can be served in the same way. The roots of the tuberous-rooted Siberian species are used as food; and perhaps those of our Spring-Beauty may be available for the same purpose. It might be worth while to select some of the larger roots, and try to improve them in size. In like manner the florist might improve the race in color by selection. The most common variety has purely white petals, but rosy shades are also abundant. Says Bryant, in allusion to the delicate color of our flower, —

> "And the Spring Beauty boasts no tenderer streak
> Than the soft red on many a youthful cheek."

So deep a rose as that shown in our plate is not, however, often met with, and we selected this specimen more especially for the purpose of showing how much nature has already done, and as an encouragement for art to do more. Our plant also varies in other respects, according to the latitude in which it grows. Nuttall notices that the leaves become more spathulate on the right bank of the Ohio; and Don remarks that they become shorter and broader as we approach Alaska. When variations are found in nature, florists may always look upon them as hints to take up improvements where nature leaves off. The Alaskan form noticed by Don may, however, belong to a closely allied species, for the botanists have not as yet determined the exact geographical range of our own.

The flowers of *Claytonia Virginica* all turn in one direction on the flower-stalk, or, as botanists say, they are secund. They expand early in the morning, but close at night. If the flowers be closely watched in these movements, it will be seen that the anthers shed their pollen on the petals, and that, when the petals are drawn in at night, they brush against the stigma, and deposit on it the pollen received from the anthers. This would be regarded by some as an arrangement for insuring self-fertilization. On the other hand, Mr. Wheeler, in the "Botanical Gazette," reports that he has noticed a tendency to heteromorphism, by which term botanists understand the occurrence of a variety of forms in the flowers of one and the same species. In some cases the pistils are proportionately longer, and in others shorter than the stamens; and in view of some experiments made by Mr. Darwin and others on primroses, this is believed to be an arrangement in favor of cross-fertilization. Hermann Müller believes that many flowers enjoy a double advantage in this respect, being so constructed that they can receive their own pollen, in case the supply, which they were originally intended to receive from another plant, should fail. It is not for us to say here whether these views — any or all of them — are wholly unobjectionable. Our chief object in these pages is not to discuss theories, but to inform the reader of all that has been learned about the plants we introduce to him, and to direct his attention to matters which may be likely to interest him.

We have ourselves noticed that in some seasons only the two lower flowers mature seed, and the failure of the others to be reproductive may have some relation to the heteromorphic condition reported by Mr. Wheeler.

The specimen from which our drawing was made came from Pennsylvania.

EXPLANATION OF THE PLATE. — 1. Complete plant, with the bulb or corm. — 2. Capsule, with a portion cut away, showing the position and small number of the seeds. — 3. Vertical section of seed-vessel, showing its triangular form. — 4. Mature seed.

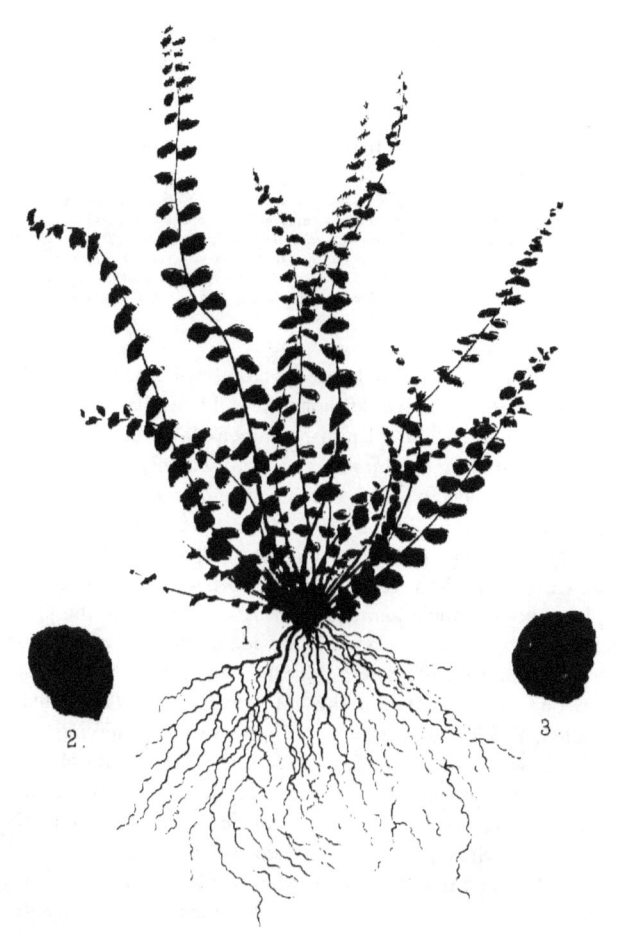

ASPLENIUM TRICHOMANES.

ENGLISH MAIDEN-HAIR; DWARF SPLEENWORT.

NATURAL ORDER, FILICES. (POLYPODIACEÆ.)

ASPLENIUM TRICHOMANES, Linnæus. — Frond pinnate; leaflets roundish, subsessile, small, roundish-obovate, obtusely cuneate and entire at base, crenate above; stipe black and polished; frond three to six inches high, lance-linear in outline, with eight to twelve pairs of roundish, sessile leaflets, three to four lines long; fruit in several linear-oblong, finally roundish sori on each leaflet, placed oblique to the mid-vein. A small and delicate fern, forming tufts on shady rocks. (Wood's *Class-Book of Botany*. See also Gray's *Manual of the Botany of the Northern United States*.)

THIS plant is not only a native of the United States, but is quite as much at home in Old England, to whose people it was known as "English Maiden-Hair," at a time when botany was still in its infancy, and had scarcely grown to the dignity of a science. The specific designation of "English" was applied to it to distinguish it from the *Adiantum Capillus Veneris*, which was called the "True Maiden-Hair." Even in those early times, however, Latin and Greek names were given to plants besides their common names; and whenever the plants mentioned by the ancients could be identified, the appellations used by them were adopted. But when no ancient name existed a new one was created, and thus our pretty little fern came to be called *Trichomanes*, from two Greek words, signifying "soft hair." It was also named *Capillaris* and *Filicula*, both of which words convey a somewhat similar meaning, but *Trichomanes* carried the day against them. Linnæus, therefore, found this name in use when he commenced to build up modern botany. But at the same time he found other ferns, which were called *Asplenium*, or Spleenwort; and as he conceived *Trichomanes* to belong to the genus which he looked upon as the true *Asplenium*, he accordingly classed the two together, and retained the former proper name of our plant as its specific appellation. This explains why the specific name in *Asplenium Trichomanes*, which stands in place of an adjective, is written with a capital. As a rule, of course, all specific names are written with a small letter; but this rule suffers an exception whenever the specific is a proper name, or is derived from one.

The reason for applying the term *Trichomanes*, or soft hair, to our plant, does not seem to be clearly established. Modern authors find in this term an allusion to the delicate, black, shining stipes (or stalks). But an old writer seems to derive it from the small, hard, black, fibrous or thready roots; and he makes this all the more probable by the manner in which he speaks of the True Maiden-Hair. This, he says, " has a root which consists of a number of blackish-brown fibres or threads, from whence springs up a small, low herb, not above a span high, whose stalks are smaller, finer, redder, and more shining than those of the *Trichomanes*."

Asplenium, as we have seen, is likewise an old name, and used to be applied to a class of plants which were held to be specifics in diseases of the spleen. In bygone times the shape of a leaf was believed to indicate its usefulness. Thus a heart-shaped leaf was supposed to be a curative in heart diseases; one that was kidney-shaped, in diseases of the kidneys; and the fact that the segments of the fronds of some of these ferns somewhat resembled the shape of the spleen, seems to have been the only reason for ascribing to them their presumed medicinal virtues. A very slender foundation, no doubt! Still, these things were as firmly believed by our forefathers as other medical matters are believed by ourselves, and perhaps with no more reason. Dr. Prior quotes an old writer as saying that, " if the asse be oppressed with melancholy, he eates of this herbe, Asplenion or Miltwaste, and so eases himself of the swelling of the spleen." He also quotes the Roman architect Vitruvius, who, in the fourth chapter of the first of his "Ten Books on Architecture," when discussing the advantages and disadvantages of the sites to be selected for cities, says that the physicians of his time cured diseases of the spleen by means of *Asplenium*, because it was found that the sheep on one side of the river Pothereus, in the island of Crete, where this herb grows, had smaller spleens than those on the other side, where it does not grow. This does not, indeed, refer to our present species, but is in place, as explaining the origin of the name *Asplenium*, which is derived from the Greek *a*, privative, and *splen*, the spleen.

Although most of the "virtues" formerly attributed to *Asplenium* were, as we have seen, mere fancies, our present species is, nevertheless, not without some merit. Syrup of Capillaire is very popular in some countries, and is said to be of real service in coughs and thoracic diseases. According to some English writers this syrup is made of our plant, although Dr. Lindley says that "Capillaire is prepared from the *Adiantum Capillus Veneris*, a plant which is considered undoubtedly pectoral, and slightly astringent, though its decoction, if strong, according to Ainslie, is a certain emetic."

The English Maiden-Hair is as nearly cosmopolitan as any species may well be. It is usually found growing in the crevices of damp, shady rocks; and according to Mr. J. H. Redfield, one of the best authorities on American ferns, the possibility of finding such a situation is the only condition which limits its distribution all over the world. Some English authorities, indeed, speak of it as occasionally growing on damp earth in shady places; but as a rule, old damp walls, or cold, shaded rocks are given as its place of abode by all the writers who treat of it. It is so easily found that few authors think it worth while to give any special locality for it. Prof. Wood is entirely silent in regard to the matter; Dr. Gray simply says "common"; Darlington speaks of it as frequent "on shady rocks and banks"; and only Dr. Chapman, in his "Flora of the Southern United States," limits it by "rocks along the

Alleghanies and northward." Dr. Haskins, in the " Botanical Gazette," reports having gathered it in Grayson County, Ky.; and collectors in New York, New Jersey, Ohio, and especially in Michigan, speak of it as abundant. Mr. Brandegee collected it in Southern Colorado; and in short, as Mr. Redfield observes, it may appear wherever the conditions are favorable. It was not found by the early botanists in the District of Columbia, as appears by Brereton's catalogue, but is now included in the list of the Potomac Naturalists' Club.

Although so common, the *Asplenium Trichomanes* is, in our estimation, one of the most delicately beautiful of all ferns. The single leaflets are, indeed, rather formal in outline, but their heaviness is relieved by the prominent veins on the upper surface, which give to them somewhat the appearance of being plaited. The contrast between the leaflets and the slender stipes is also very attractive, and calls up the idea of weakness and strength happily united. There is, moreover, a great deal of intellectual pleasure to be derived from seeing this little plant growing in its native locations. Many of our gay-flowering plants will only deign to exhibit their charms in a very limited circle of high society, where they are petted and pampered. But this little fern, like a good angel, goes forth over the wide world, seeking out the cold, cheerless spots which are despised and left in utter loneliness by its gayly colored companions, and decks them with an elegant and chaste beauty which even the more aristocratic members of the floral kingdom might envy. If any poet wishes to find an emblem of universal love, and of charity to the poor and forsaken, he cannot certainly choose anything better befitting the idea than our English Maiden-Hair.

EXPLANATION OF THE PLATE.— 1. Complete plant from a specimen gathered in Massachusetts. — 2. Leaflet enlarged, showing upper surface. — 3. Leaflet enlarged, showing lower surface and sporangia.

ANEMONE CAROLINIANA.

CAROLINA ANEMONE.

NATURAL ORDER, RANUNCULACEÆ.

ANEMONE CAROLINIANA, Walter. — Stem slender, one-flowered ; peduncle many times longer than the small, sessile, three-leaved, three-toothed involucre ; radical leaves two to three, long-petioled, ternate, deeply parted, lobed and toothed ; sepals fourteen to twenty, oblong, white ; achenia numerous in a cylindrical-oblong head, woolly; stems six to twelve inches high ; flowers one inch in diameter. (Chapman's *Flora of the Southern United States*. See also Gray's *Manual of the Botany of the Northern United States*, Wood's *Class-Book of Botany*, and Torrey and Gray's *Flora of North America*.)

THE reader who will carefully compare Dr. Chapman's description with our plate cannot fail to be startled by the discrepancy in regard to the color of the flower; for while he distinctly and unequivocally states the color to be *white*, our illustration as unmistakably shows it to be *violet* or *purple*. The discrepancy will be readily understood, however, by those who are accustomed to deal with flowers. Variations in color are frequently found, and the *Anemone Caroliniana* affords a good example. This also explains why the various authors differ so widely in speaking of the flower. Thus, Dr. Gray says, "purple or whitish"; Prof. Wood, "white or rose-colored, . . . outer sepals dotted with purple"; and Torrey and Gray, "white, often tinged or spotted with purple." The beautiful specimen from which our drawing was made, and which was kindly sent to us from Western Kansas by Mr. Sternberg, makes it evident that still another must be added to this list of variations, namely, violet or purplish.

The *Anemone* is frequently mentioned in ancient Greek and Roman mythology and poetry; but from the attending circumstances, it seems that the various stories with which the name is connected relate to different species of the genus. The sad tale, for instance, of the fair maiden who fell in love with Zephyr, and was banished from her court by Flora, and finally destroyed by the rude blasts of Boreas, seems to be most in accord with the character of the Wind-Flower or *Anemone nemorosa;* while the flower which is involved in the story of Venus and Adonis must evidently have been more brilliant in color, and somewhat like our own *Anemone Caroliniana.* It is well known that Venus, or Aphrodite, as the Greeks called her, was enamored of a beautiful mortal, a youth named Adonis, and that when Adonis had been killed by a wild boar while hunting, Venus caused flowers to spring up out of the blood of her lover. This version of the creation of *Anemone* is related by Ovid, the celebrated Roman poet, in the tenth book of his "Metamorphoses," and has been translated into English by Eusden as follows: —

> "'For thee, lost youth, my tears and restless pain
> Shall in immortal monuments remain;
> With solemn pomp, in annual rites return'd,
> Be thou forever, my Adonis, mourn'd.
> Could Pluto's queen with jealous fury storm,
> And Menthe to a fragrant herb transform,
> Yet dares not Venus with a change surprise,
> And in a flower bid her fallen hero rise?'
> Then on the blood sweet nectar she bestows;
> The scented blood in little bubbles rose, —
> Little as rainy drops which fluttering fly,
> Borne by the winds, along a lowering sky.
> Short time ensued till, where the blood was shed,
> A flower began to rear its purple head.
>   .   .   .   .
> Still here the fate of lovely forms we see,
> So sudden fades the sweet anemone."

Shakespeare also, in his poem entitled "Venus and Adonis," mentions this myth: —

> "By this, the boy that by her side lay killed
> Was melted like a vapor from her sight;
> And in his blood, that on the ground lay spilled,
> A purple flower sprang up, chequered with white."

It will be noticed that Ovid's "purple head" agrees quite well with the variety represented by our drawing, while Shakespeare's "purple, chequered with white," answers tolerably well to Torrey and Gray's "white, spotted with purple."

We must, however, break through the spell of these poetical illusions, which have carried us far away to the sunny lands of Greece and Italy, and return to the truth of reality by remembering that our *Anemone Caroliniana* cannot be absolutely identical with the flower born from the blood of Adonis, as it is specifically American. It differs from many of its kindred also in its places of growth; for while some *Anemones* prefer to grow in the recesses of deep forests, and while the delicate *A. nemorosa* seeks the shade of scattered woods, the *Caroliniana* delights in open places, and in the full blaze of the western sun. If we may be permitted once more to indulge in a poetical revery, we might almost imagine our flower as fleeing from the dangerous localities which, in their cosy seclusion, are so well fitted to be the abodes of lovers, and seeking the broad light of day, so as to avoid the sad fate which befell her unfortunate sister, whom she had seen —

> "Loving with all the wild devotion,
> That deep and passionate emotion ;
> Loving with all the snow-white truth
> That is found but in early youth ;
> Freshness of feeling, as of flower,
> That lives not more than spring's first hour."

We have just said that the Carolina Anemone delights in the blaze of the western sun, and indeed its geographical range extends across the plains to the base of the Rocky Mountains, and thence takes a southerly course (if Mr. Watson's view be adopted, that it is the same as *A. decapetala*), through Utah, Arizona, and New Mexico, to Peru, Chili, and Brazil. This shows that its geographical centre must be to the south, while the centres of most other common kinds of Anemone lie towards the north. It also becomes evident from this that the name of the plant is not very aptly chosen, as its range extends far beyond the limits of the Carolinas.

The root-structure of our species is worthy of a more complete study than we have been able to give to it. As far as we can ascertain, the travelling rhizoma, or rootstock, produces a succession of small tubers, which throw up leaves, or leaves and flowers, the season following that in which they were produced. Generally, the tuber is formed by the thickening of the end of the rhizoma, as in the potato. A rhizoma is really a stem, with this difference only,—that instead of growing above, it grows under ground. In the case of the potato, the thread-like growth of the rootstock as soon as it has advanced six inches or so from the parent stem, thickens, and forms a tuber, which we call a potato; but occasionally this tuber will start a new growth from its apex the same season, which again thickens at its end, and from this second tuber even a third rootstock sometimes strikes out, which also forms a potato at its end, so that finally the whole assumes something of the shape of a necklace, or of large beads strung upon a string at certain intervals, the end, however, being always a tuber. Our *Anemone* grows in the same way. On the right-hand side of our drawing we see the remains of the rootstock growth of last year, which connected with the plant of that season. This, we believe, dies at the end of the year. We see, also, that after making one small tuber our plant started to make another, and as this second was stronger than the first, it was able to make three flowers, while the first had but one. On the left, we have the growth made since the last year's tubers threw up their leaves and flowers, and this new rootstock is also thickening for a tuber for the next season.

The Carolina Anemone, if we may judge from its western location, in a hot, dry region, will be very well adapted to garden culture. In our own garden, it has taken good care of itself for two years; and its bright, purple flowers, opening before the first of May, among the many white flowers of that season, render its presence in the garden-border very desirable indeed.

ROSA CAROLINA.

SWAMP ROSE.

NATURAL ORDER, ROSACEÆ.

ROSA CAROLINA. Linnæus. — Stem erect, smooth, armed with stout, recurved, stipular prickles; leaflets five to nine, oblong or elliptical, acute, finely serrate, dull and smoothish above, the lower surface paler, or, like the prickly petioles and caudate calyx-lobes, tomentose; flowers single or corymbose; calyx-tube and peduncles glandular-hispid; stem four to six feet high, commonly purplish; fruit depressed-globose, glandular. (Chapman's *Flora of the Southern United States.* See also Gray's *Manual of the Botany of the Northern United States,* and Wood's *Class-Book of Botany.*)

THE botanists of the earlier part of this century frequently gave specific names to mere varieties, since they were not as well informed as those of our own time in regard to the tendency to variation in plants and flowers, a tendency which is shown much more distinctly in some species than in others. But later, when it was not thought necessary to specially note these variations, their names, previously given, often remained as synonyms to burden botanical nomenclature; and hence the greater the tendency to vary, the more synonyms a plant may have. Our *Rosa Carolina,* being a very variable species, furnishes a good illustration of this statement. From the list of synonyms given by Mr. Watson in his "Bibliographical Index to North American Botany" we select the following as of most importance: *R. Virginiana,* by Du Roi; *R. corymbosa,* by Ehrhart; *R. Caroliniensis* and *R. palustris,* by Marshall; *R. Pennsylvanica,* by Michaux; *R. florida,* by Don; *R. flexuosa* and *R. cuneaphylla,* by Rafinesque; and *R. Hudsonica,* by Thory. Several of these names show that they were based on the number of flowers in a cluster, or of leaflets in the leaf, or on other peculiarities which are now known to be of little consequence in the Rose, although they may perhaps be of some weight in other genera.

Of all our native species, the *Rosa Carolina* is perhaps the most variable, not only as a garden plant, but even in its wild state. Like other Roses in their natural condition, it has normally only five petals; but flowers with a larger number are not unfrequently found, and Humphrey Marshall, in his "Arbustum Americanum," published in 1785, describes a perfectly double Rose, which seems to be identical with our species, although he calls it *R. Pennsylvanica plena*. Rafinesque, indeed, seems to have found several double forms. He notes not less than seven different varieties, to which, in accordance with the custom of his day, he gave Latin varietal names, such as *corymbosa, uniflora, alba, erecta,* and *pimpinellifolia*, which latter, he says, may have single or double flowers, and very small leaves. He adds by way of conclusion: "There are many varieties, several of which have produced double flowers in gardens." These varieties were, no doubt, first discovered in a wild state, and then transplanted to the garden, although our author states that they are found in cultivation.

Among the many varieties mentioned by Rafinesque, the white one (*alba*) is especially interesting in connection with the legendary history of the Rose. From the various stories of the birth of this flower, it is evident that the original Rose was conceived to be white, and that the colored varieties were looked upon as a departure from the state of nature. This idea is embodied in the following lines by one of the poets: —

> "As erst, in Eden's blissful bowers,
> Young Eve surveyed her countless flowers,
> An opening rose of purest white
> She marked with eye that beamed delight.
> Its leaves she kissed, and straight it drew
> From beauty's lip the vermil hue."

But whatever may have been the fact in regard to the Roses of the primeval world, it is nevertheless true that, among our native Roses, color is the rule and white the variety; and the latter is indeed so scarce that we know of no author but Rafinesque who refers to it. The white varieties seem to have disappeared even from cultivation, as we have met with no one who has seen any of late years.

The Swamp Rose has not the grateful perfume of the Dwarf Wild Rose, nor has it the perfectly outlined petals, or the classical look in general, of that species. The flowers, indeed, have a somewhat loose and ragged appearance; but the plant, nevertheless, presents certain features which delight the eye. It is generally found growing in large numbers together, often covering a whole acre or so; and in June, when the bushes are in their flowering prime, the mass of blossoms is beautiful to look upon; while in the autumn, when the leaves of our plant are of an orange brown, and all the bushes are aglow with the crimson, berry-like fruit, there is hardly a more attractive sight to be seen. The height of the Swamp Rose is about twice that of the Dwarf Wild Rose, and the peculiar gray of the under surface of the leaves, together with the dull, dark green of the upper surface, affords a good mark of distinction. The two will seldom, indeed, be confounded by the careful student, no matter how much the Swamp Rose may vary from its original form; but if there should be any difficulty in determining the species, the spines will decide the question, as they are straight in the Dwarf Wild Rose, and hooked in the present species. In the latter, the calyx leaves also remain on the fruit much longer than in the former, but they fall completely before winter sets in. On the specimen represented in Fig. 2 of our plate, they are still partly to be seen; and we may here remark that, while botanical authors speak of the fruit as "depressed-globose," our drawing is a faithful representation from nature.

The attention of the poets has, so far, been given to the Roses of the Old World almost entirely; and indeed the only direct poetical allusion to any of our native species that we can find is by Mrs. Sarah J. Hale, who makes our Swamp Rose the emblem of dangerous love in the language of flowers. In general character, our species approaches very near to the *R. cinnamomea*, or Cinnamon Rose of Europe, of which there is a thornless variety, and the Swamp Rose is also frequently found very sparingly armed. In that interesting book entitled "Legends of the Rose," we are, indeed, told that all Roses were originally thornless, and the flower itself is thus made to explain the existence of the thorns: —

> "Young Love, rambling through the wood,
> Found me in my solitude,
> Bright with dew and freshly blown,
> And trembling to the zephyr's sighs;
> But as he stooped to gaze upon
> The living gem with raptured eyes,
> It chanced a bee was busy there,
> Searching for its fragrant fare;
> And Cupid, stooping too, to sip,
> The angry insect stung his lip,
> And gushing from the ambrosial cell,
> One bright drop on my bosom fell.
> Weeping to his mother, he
> Told the tale of treachery;
> And she, her vengeful boy to please,
> Strung his bow with captive bees;
> But placed upon my slender stem,
> The poisoned stings she plucked from them,
> And none, since that eventful morn,
> Have found the flowers without a thorn."

The Swamp Rose is at home along the seaboard, from Maine to Florida, but beyond the Mississippi it occurs only in Iowa and in the eastern part of Nebraska, and as far as we know, it has never yet been found either directly north or south of these states.

EXPLANATION OF THE PLATE. — 1. Flowering branch from a Massachusetts specimen gathered toward the end of June. — 2. Fruit from Pennsylvania in October.

PACHYSTIMA CANBYI.

CANBY'S MOUNTAIN-LOVER.

NATURAL ORDER, CELASTRACEÆ.

PACHYSTIMA CANBYI, Gray.—Surculosely creeping; leaves oblong-linear, slightly denticulate; pedicel filiform, elongated; petals oblong-ovate; style very short. (Gray, in *Proceedings of the American Academy of Arts and Sciences*.)

THE plant which we are about to introduce to our readers is one of those which do not attract by showy flowers. Nevertheless, the rich hue of its evergreen leaves gives it a unique character among our native plants, and will make it valuable in the eyes of those who love beautiful foliage, while readers of a more scientific turn of mind will find much of interest in its family history. The genus *Pachystima* consists of only two species, and was not known until the celebrated expedition made across the continent by Lewis and Clarke in the years 1803-1806, when specimens of one of the species were brought home from the Rocky Mountains by Lewis. Pursh thought it was a holly, and so named it *Ilex Myrsinites*. Nuttall, with probably better specimens before him than Pursh, made it out to be a *Myginda*, which was much nearer the truth, as *Myginda* is a genus of the order *Celastraceæ*, to which our plant belongs. Later, Nuttall himself discovered essential differences between *Myginda* and the plant originally discovered by Lewis, and established the latter as a new genus, under the name of *Oreophila*. Rafinesque, however, had already discovered the distinction, and had named the genus *Pachystima* before Nuttall published his name, and Rafinesque's name, therefore, was generally adopted, in accordance with the ethics of botany, which demand that the name first published with a description showing the distinctive character of the plant to which it is applied shall have precedence.

The derivation of Rafinesque's generic name, *Pachystima*, is not clear. The pedicel, or flower-stalk, is filiform, as given in Dr. Gray's description, but thickens just beneath the receptacle in both the species belonging to the genus, and if the "thickness," which the name implies, refers to this feature, it would seem to be appropriate. Rafinesque adopted Pursh's specific name for the only species then known, and thus we had *Pachystima Myrsinites*. In Torrey and Gray's "Flora of the United States" it is, however, described under Nuttall's manuscript name of *Oreophila myrtifolia*, or Myrtle-leaved Mountain-Lover, in allusion both to the character of its foliage and its home in the mountains. This species, the *P. Myrsinites*, has since been found in many of the mountain localities in the Northwest and in British North America.

The second species, *Pachystima Canbyi*, was not discovered till 1858, when it was seen by Mr. William M. Canby, of Wilmington, Del., on a bluff along the New River, near White Sulphur Springs, Va.; but it was only in 1868 that Mr. Canby was able to procure good specimens, from which Dr. Gray described and named the plant. Subsequently, our species has been collected in several other places in Virginia by Mr. Howard Shriver, and it is quite likely that it will be found not uncommon along the great Alleghany ridge.

The order *Celastraceæ*, to which *Pachystima* belongs, is nearly allied to the *Rhamnaceæ*, or buckthorns, but differs from them in several particulars, the most characteristic being that the stamens in the latter are always opposite the petals, provided these are present; while in the former they are alternate with them, as shown in our enlarged flower (Fig. 3), where they are seen fronting the larger sepals, the smaller, oblong-ovate petals lying between. *Celastraceæ* itself is not a very large order, but is, nevertheless, tolerably well known to most persons from the *Euonymus*, familiarly called the "Spindle-Tree" or "Burning-Bush." The order is again divided into two general sections, the one to which the *Euonymus* belongs having a rather dry capsule, opening to let out the somewhat fleshy seed; while the other, in which our *Pachystima* is placed, has drupacious fruit, or in plain English, a kind of fruit which resembles stone-fruit. With the easily obtained *Euonymus* before him, the student can readily gain a fair idea of the two divisions of the order. The berries on our species, however, seem to be sparingly produced, and the only ones we ever saw were in a dry condition on Mr. Canby's specimens. Although the plant from which the accompanying plate was drawn has flowered freely in cultivation for several years, it has never produced any fruit; but as in the case of the flowers themselves, it is not likely that the berries would add much beauty to our pretty evergreen plant.

The fact just alluded to, that the *Pachystima Canbyi* produces berries but sparingly, opens up a question which was already discussed by the botanists of the preceding generation, in connection with the sister species, *P. Myrsinites*. The question is, whether the plant may not prove to be, in many cases, monœcious, or even, practically at least, diœcious. Nuttall believed *P. Myrsinites* to be monœcious, or having the male organs in one set of flowers, and the female organs in another. Torrey and Gray, on the contrary, thought it must be hermaphrodite, or with both kinds of organs in each flower, more especially so as Sir William Hooker had figured it that way. But modern experience shows us that even when both kinds of organs are apparently perfect, the one or the other may be defective, and hence the plant may be practically monœcious, or indeed even diœcious, if it should so happen that on some individuals all the male organs are defective, and on others all the female organs. The flowers on our plant seem perfect, but, as already stated, produce no fruit.

The plant increases by branches running under the ground,

and rooting, if the soil be light, or by sending out roots from branches that find themselves near the ground, or covered by loose vegetable matter. The early spring-shoots have the leaves very variable in form, from linear to ovate, and much more sharply denticulate than those which appear on a second growth of branches, sent out later in the season. It is on these later branches that the flowers appear in the following spring.

To cultivators the plant will prove very acceptable as an evergreen dwarf bush. In the writer's garden it has a frame, a shallow, bottomless box, a few inches deep, placed around it, filled with sand, into which it seems to love to root. The rooted pieces are easily transplanted to form other colonies. Little pieces of cuttings also root very well in pans of sand, set in an ordinary green-house.

The fact that a distinct genus like *Pachystima* should have only two representatives, and these confined to limited areas over this great continent, will be a subject of speculation with those interested in the genesis of plant-forms. Are these species new forms, which have appeared comparatively quite recently, and which by and by will become more numerous by developing into varieties and other species, or are they very old forms, now in process of extinction? The time may come when there will be circumstantial evidence sufficient to answer these questions, and the earnest attention which they command among scientific men at the present time springs from the belief that it will eventually be possible to answer them satisfactorily.

Our plant has absolutely no common name, and by way of rectifying this omission we have ventured to call it "Canby's Mountain-Lover," for reasons which must have become apparent to the reader in the course of this article.

EXPLANATION OF THE PLATE.—1. Main branch, with secondary branches, showing the denticulation of the leaves.—2. Branchlet of the second growth, with entire leaves, in flower in spring.—3. Flower magnified, showing the position of the anthers, and the symmetrical arrangement of all the parts.

SPIRANTHES CERNUA.

DROOPING-FLOWERED LADIES' TRACES.

NATURAL ORDER, ORCHIDACEÆ.

SPIRANTHES CERNUA, Richard. — Stem leafy below and leafy bracted above, six to twenty inches high; leaves linear-lanceolate, the lowest elongated, four to twelve inches long, two to four lines wide; spike cylindrical, rather dense, two to five inches long, and with the flowers either pubescent or nearly smooth; perianth horizontal or recurving, the lower sepals not upturned or connivent with the upper; lip oblong and very obtuse when outspread, but conduplicate or the margins much incurved, wavy-crisped above the middle, especially at the flattish and recurved-spreading apex, the callosities at the base prominent, nipple-shaped, somewhat hairy; gland of the stigma linear, in a long and very slender beak. (Gray's *Manual of the Botany of the Northern United States.* See also Wood's *Class-Book of Botany*.)

THE plants now called *Spiranthes* were placed in the genus *Ophrys* by Linnæus, and in that of *Neottia* by his contemporary Willdenow, and under the names of these genera they must be looked for by the historical investigator. Our own botanist Nuttall, in his earlier works, classes them with *Neottia*, but in his later writings (1827), he calls them "*Spiranthes*, a section of the genus *Neottia*." The genera *Ophrys* and *Neottia* still exist, and have given their names to two of the various tribes into which the order *Orchidaceæ* is divided; but *Spiranthes* has now been universally adopted as the generic name of the plants to which our species belongs, even by English authors, with whom the old *Neottia spiralis* is at present *Spiranthes autumnalis*. This last-named plant is probably the only representative of the genus in England, nearly all of the fifty species or so which compose it being natives of the New World, although only a very few of them are found within the limits of the United States.

Among our American genera there are two, besides *Spiranthes*, which belong to the tribe *Neotteæ*, namely, *Goodyera* and *Listera*, *Spiranthes* being intermediate between them. All the species of the genus *Spiranthes* have a callous protuberance at the base on each side of the lip, while those of the other two genera have none; *Listera* has all the sepals and petals spreading, and thus differs from its fellows, the petals of the latter being so arranged as to be ringent (or gaping) at the base. Many other peculiarities of more or less importance might be pointed out as characteristic of the different genera, but it is hardly necessary to do this, as there is seldom any difficulty in determining the genus of these plants from their general appearance. The species, on the contrary, are very difficult of determination, as there are many varieties of each, which, by their great apparent differences, are calculated to puzzle the student. Of *S. cernua*, for instance, according to Dr. Gray, the commoner form has pure white, sweet-scented flowers, grows in wet places, and often loses nearly all its root-leaves at flowering-time, while one variety grows in dry ground, has greenish, cream-colored, stronger-scented flowers, and retains its root-leaves.

The old name *Neottia* is Greek for bird's nest, and was given to our plants, says an old writer, "because the plaiting of the roots one among another resembled a crow's nest." *Spiranthes* is also from the Greek, *speira* meaning a spiral or coil, and *anthos* a flower, and seems to have been suggested by the apparently twisted arrangement of the flowers, which strikes every observer. The old English name was "Ladies' Traces," from the resemblance of the twisted spikes to the silken cords or laces, formerly called "traces," with which fair dames used to gird themselves and fasten their various articles of dress before hooks and eyes, buttons, pins, and the like were invented. The word has become almost obsolete in this connection now, being applied only to the cords or ropes by which horses are attached to the plough, or to the leather straps of more pretentious harness. The original meaning of the word having thus been forgotten, modern authors spell the name of our plant "tresses," and suppose it to have been adopted from the resemblance to a tress or curl of hair; and perhaps the two words may originally have been derived from one root, for certainly many flowing tresses have proved to be the traces by which masculine hearts were chained to the triumphal car of beauty.

The specific name, *cernua*, is from the Latin, and alludes to the habit which the flowers have of turning their faces downward. *Spiranthes cernua* might, therefore, be called in English "Drooping-flowered Ladies' Traces."

We have already noted that there are many varieties of the different species of Ladies' Traces; and in view of this fact, it will be best for the student to consider all the characteristics very carefully in trying to determine a species, and then to strike an average from the whole. There are several points, however, which will materially assist the young botanist. The first of these is the division of the genus into two sections. According to Dr. Gray, the species in one of these sections have the flowers in *three* ranks, crowded in a close spike, while those in the second have the flowers in *one* straight or spirally twisted rank. In the latter case, we may picture the arrangement of the flowers to ourselves if we imagine them set upon a string, and this string wound in a spiral around a stick; in the first case, there are three such strings running closely parallel to each other, and also twisted round the stick as before indicated. Our present species belongs to this three-ranked division. Dr. Chapman and Prof. Wood have essentially the same arrangement. The roots also offer some good specific characters, being a mass of fleshy fibres in some species, as in *Spiranthes cernua* (Fig. 3), and quite tuber-like in others. In some cases, again, the rachis, that is to say, that part of the stem to which the flowers are attached, is perfectly straight, and only the flowers seem coiled around it, while in other species it is screw-like, and seems to carry the flowers with it as it coils.

The interest which the orchidaceous plants have always inspired has been considerably increased by the publication of Mr. Darwin's writings, and more especially by his work on the "Fertilization of Orchids." In this book, the celebrated evolutionist devotes considerable space to the genus *Spiranthes*, and also mentions our species, *S. cernua*. After a thorough discussion of the matter, he comes to the conclusion that everything in these plants is most beautifully contrived so "that the pollinia should be withdrawn by insects visiting the flowers"; and finally closes his remarks with the following sentence: "Then, as soon as the bee arrives at the summit of the spike, she will withdraw fresh pollinia, will fly to the flowers on another plant, and fertilize them, and thus, as she goes her rounds and adds to her store of honey, she continually fertilizes fresh flowers, and perpetuates the race of our autumnal *Spiranthes*, which will yield honey to future generations of bees."

It is very singular that a plant with such a suggestive common name should never have attracted the attention of the poets; and yet this seems to be the case, at least as far as our own reading extends. The fact appears still more remarkable when we consider the delicious fragrance of the flower and the peculiar circumstances under which it is found, often growing entirely alone, far away from its orchidaceous relations, and coming into flower long after most of the family have betaken themselves to rest.

Our species seems to be confined to the Eastern States, and to be found very seldom, if ever, beyond the Mississippi; but, like some of the other members of the same family, it generally grows in considerable quantities wherever it does occur.

EXPLANATION OF THE PLATE. — 1. Flower-scape.— 2. Central portion of scape, showing the sudden arrestation of leaves and their transformation into bracts.— 3. Root.

PHLOX REPTANS.

CRAWLING PHLOX.

NATURAL ORDER, POLEMONIACEÆ.

PHLOX REPTANS, Michaux. — Stem erect, with procumbent runners at the base bearing roundish-obovate and rather fleshy subsessile leaves; upper stem-leaves ovate-lanceolate; corymb few-flowered; stem four to six or eight inches high; leaves about an inch long, more or less pilose and ciliate, — the lower ones spatulate-obovate, tapering to short margined petioles; corolla deep purplish-red, — the tube about an inch long, a little curved. (Darlington's *Flora Cestrica*. See also Gray's *Manual of the Botany of the Northern United States*, Wood's *Class-Book of Botany*, and Chapman's *Flora of the Southern United States*.)

MOST of the Phloxes of the Eastern United States were well known to the botanists of the earlier part of the present century, and the species to which this chapter is devoted was one of their latest discoveries. It was first noticed in the mountains of North Carolina by Michaux, who described it, and gave it its present name, *Phlox reptans*. Shortly afterwards the same species was also found in Georgia by John Frazer, an English collector, and a representation of the plant appeared in the "Botanical Magazine," where it was described as *Phlox stolonifera*. Frazer also sent seeds to England, from which flowering plants were produced about 1800.

This incident is well calculated to show the origin of synonyms, which are so often a source of annoyance and difficulty to the student. It must necessarily happen now and then that two people discover and describe the same thing simultaneously, or very nearly so, without having any knowledge of one another's work, or that some one describes a plant as new which is afterwards found to be different in no essential particular from one already described. In such cases the rule, that the oldest name shall have the preference, should be strictly adhered to. But our plant is still very generally called *Phlox stolonifera* by English authors, while American authors have without exception, and very justly, adopted the name *Phlox reptans*, as first used by Michaux.

It is curious to note the coincidence in the names given to our plant by Michaux, and by Curtis in the "Botanical Magazine," without any knowledge on the part of the one, of the doings of the other. For *reptans* is the Latin for "crawling," and *stolonifera* signifies "stolon-bearing," stolons being trailing or reclined and rooting shoots, or runners, which creep along the ground, like the runners of the strawberry. And indeed the peculiarity to which this species owes its distinctive appellation is very striking. Most of the Phloxes are what are called herbaceous plants; that is to say, the stalks die down to a rootstock or crown every year, and there is nothing left of the plant during winter but bud-like eyes, from which the flower-stalks and leaves push up in the spring. The *Phlox reptans*, however, is an evergreen, and the way in which it grows is well shown by our artist. The plant sends out a runner or stolon, and from the terminal bud, made at the end of the stolon in the fall, a central flower-shoot ascends, together with another shoot which bears nothing but leaves. Besides these two shoots, however,— both of which die in the fall, the leaf-bearing one seemingly without having accomplished anything,— a number of others push up, some of which are only scantily clothed with leaves, while the rest bear no leaves at all. The scantily leaved shoots often root at the tip, but the best plants for the future are produced by the leafless runners, which form a bud at the end with roots, and then die. All these various kinds of shoots can be seen in our drawing. In the middle is the flower-stem, to its left is one of the scantily leaved shoots, to the right the full-leaved shoot, and part of the leafless, creeping runner, which is destined to form a good, strong, new plant. In the spring, when growth commences, small fibres push out from the old runner (a feature which may also be observed in our plate), which thus helps to sustain the plant, but the next year all this dies away. The plant is in reality a wanderer; and in culture, to which it readily adapts itself, it has to be watched, and must now and then be brought back to its proper quarters, as otherwise there is danger that it will quietly walk away and eventually disappear entirely from the florist's collection.

It seems almost unnecessary to call attention to the advantages which this plant offers to the ornamental designer. The almost entire leaves, of a noble simplicity of form; the very straight and precise flower-stalk; the few flowers, set on the summit, at regular distances from each other, like the arms of a candelabrum; the corolla, with its rounded segments arranged carefully one over the other, and disposed so as to produce a symmetrical outline; and finally the whole arrangement of the parts in their relation to each other, — all these go to make a combination which can readily be turned to good use where more graceful lines would not be in harmony with the surroundings. In cases where an expression of strength is desired, our plant might be excellently well employed ornamentally, to emphasize the functions of the constructive parts; and as it is strictly an American plant — a member of an exclusively American family — it will be appropriate in connection with any work of a national character.

The geographical distribution of this beautiful Phlox has not been fixed as definitely as it might have been by this time, considering that it is limited to the older settled portions of our own country. We have already seen that it was found in North Carolina by Michaux, and in Georgia by Frazer. Drummond is credited with having found it "in the Alleghanies," but this certainly is not very definite. Mr. Peters is cited as an authority for the statement that it exists in Kentucky, and Prof. Wood says, "hillsides and mountains, Indiana to South Carolina." Dr. Chapman says, " damp, shady woods near Washington, Wilkes County, Georgia, and northward along the mountains"; and

Dr. Gray, finally, gives in a general way, "damp woods, Pennsylvania, Kentucky, and southward." The local Floras rarely mention it. It is not in Willis' catalogue of New Jersey plants, nor in Beardslee's list for Ohio. Coleman reports it as occurring in Michigan, and in the "Botanical Gazette" he speaks of a white variety which he found at Grand Rapids, Iowa. We have seen that Prof. Wood gives Indiana as one of its locations, but it does not occur in any of the counties of which the Floras are given in the Geological Survey of that State. It evidently prefers to keep to high elevations, chiefly in southern ranges, and there it will probably be found most at home in cool, moist woods.

The time of flowering of the *Phlox reptans* is given as May by some authors, and as June by others; while English writers, who, of course, speak of the plant only in its cultivated state, give it as from May to September. A good deal, no doubt, depends on the situation. In a warm, sunny spot its flowering time would perhaps be shortened. It does quite well in our gardens, however, and with proper attention it would probably become the parent of a very beautiful race. All the Phloxes are changeable, and this species is not likely to be an exception to the rule, as it shows some variations in color, even in its natural state. Some writers describe the flowers as rose, purple, or pale red, and a pure white variety is reported by Mr. Coleman, as before stated. Without a doubt, therefore, its capabilities for floral improvement must be very great.

As the plant is not frequent where the foot of man usually treads, it has not yet attracted general attention, and hence is still without a generally accepted common name. We must therefore be satisfied with the translation of its botanical name, "Crawling Phlox," as given by Dr. Darlington.

CHRYSOPSIS MARIANA.

MARYLAND GOLDEN STAR.

NATURAL ORDER, COMPOSITÆ (ASTERACEÆ OF LINDLEY)

CHRYSOPSIS MARIANA, Nuttall. — Perennial; stem one to two feet high, simple, covered with loose silky deciduous hairs; lowest leaves spatulate-oblong, entire or slightly serrate; the upper ones lanceolate, sessile, entire; corymb small, mostly simple and umbellate, cone-like in the bud; peduncles and involucre glandular. (Chapman's *Flora of the Southern United States*. See also Gray's *Manual of the Botany of the Northern United States*, and Wood's *Class-Book of Botany*.)

THE natural orders into which plants are divided have, with but few exceptions, been named from some representative genus belonging to them. Thus the order of *Rosaceæ* received its name from *Rosa*, or the Rose Family; *Ranunculaceæ* from *Ranunculus*, the Crowfoot Family; and so on. Among the exceptions alluded to, the order which embraces *Chrysopsis Mariana* is generally found; for it is called *Compositæ* by most botanists, not from any of the genera belonging to it, but rather as descriptive of the compound character of its flowers, each flower, although having the appearance of but a single one, being in reality composed of an aggregate of a number of florets or small flowers, all placed on one common receptacle. Dr. Lindley, however, a well-known English writer on botany, endeavored to secure uniformity in this respect, and in his book entitled "The Vegetable Kingdom," he therefore dropped the few exceptional names, and replaced them by others modelled on the general rule. To the old order *Compositæ* he gave the name of *Asteraceæ*, from the large and celebrated genus *Aster*, which belongs to it. Most modern botanists, indeed, do not seem to have adopted Dr. Lindley's views, although the reasons advanced by him are certainly very good. Under these circumstances, we have thought it advisable to give both names a place at the head of our chapter.

The genus *Chrysopsis* is not far removed from the true *Asters*, and is intermediate between them and the European genus *Inula*, to which Elecampane belongs. The species to which this chapter is devoted was itself formerly looked upon as an *Inula*, and Miller, in his "Gardener's Dictionary," published in 1760, speaks of it as *Inula Mariana*. Nuttall was the first to point out the essential differences between the two genera in 1818, and to him we also owe the present name of our genus, *Chrysopsis*, which, as he tells us, he gave to it from the fact that, although it had some of the characters of a class of Asters with corymbose inflorescence, it always differed from the latter in the "prevailing yellow color of the flowers." In most cases, botanists regard color very slightly in establishing the characters of a genus, and the peculiarity must, therefore, have been very striking in this instance to have induced Nuttall to base a generic name on it. *Chrysopsis* is from the Greek words *chrysos*, gold, and *opsis*, aspect, appearance, sight. Our genus, however, differs from *Inula* not only in general appearance, which, as a matter of course, carries some weight in a natural system, but there is also a difference in the seeds. In our plants, they are compressed and ovate-oblong, while in *Inula* they are either four-sided or round.

As *Inula* is not a real native plant in the United States, although it grows wild in many localities, having escaped from gardens, it will not find a place in our work; but considering that it was so closely connected with our *Chrysopsis* for a long time, and as we may have no opportunity to refer to it again, we may perhaps be excused for dwelling for a moment on the history of the family name, which formerly used to be common to both. The botanical name of Elecampane is *Inula Helenium*. The attempts to trace the etymology of the generic appellation, *Inula*, have been given up as barren by most botanists, but it seems to us that *Inula* may be a corruption of the original name of the plant, which, according to the earlier accounts, was probably dedicated to St. Helen by some of the Eastern nations. In Italy, also, it is often spoken of as "Elenio," and even as far north as Denmark, it is generally called "St. Helen's Rood." It is hardly necessary to point out that the same name is still preserved in the specific appellation of the Elecampane, *I. Helenium*. Bauhin, one of the oldest botanical writers, says it is probably the plant referred to in some legend as having sprung from tears shed by the famous Helen of Troy. The oldest name found in the herbals, or books on herbs, appears to be "Ala campana," which name is based on the fact that the plant is abundantly found in the Campana, the country around Naples. We see that, whatever may have been the origin of *Inula*, the derivation of the common name, Elecampane, is clearly accounted for.

The *Chrysopsis Mariana* seems to have been the first of its genus known in England, where it was introduced in 1742, according to Philip Miller, by Dr. Thomas Dale, from Maryland, whence its specific name *Mariana*. Since that time a number of other species have been discovered, both in the Atlantic and the Pacific regions of our country. Our present species is not found north of New York, but within that State it occurs in many places on Manhattan Island, and becomes very abundant in New Jersey, where it is one of the commonest plants in the dry and sandy barrens. It is not common in Pennsylvania, although not infrequent in the region drained by the Wissahickon, which supplied the specimen from which our drawing was made. Towards the West, it does not appear to extend north of Southern Ohio, but from there southward to Florida it is often met with on the lower elevations, becoming more abundant as it approaches the sea-coast. It is in no list from Kentucky, west of the mountains, as far as we know, but it probably grows in some of the Mississippi States. In Pennsylvania it is generally found in half-shaded woods, but in other states it seems to favor more open places.

In our description, as quoted from Dr. Chapman's work, we have spoken of the stem as being "covered with loose, silky, deciduous hairs," and the attentive reader may have noticed that this hardly corresponds with our drawing. The explanation of the apparent discrepancy is, however, foreshadowed in the word "deciduous." Our plant, in common with several other species of the same genus, has a peculiar, cobwebby appearance when young, which always attracts the attention of the observer. This appearance is due to the light and tangled hairs which then clothe the stem, but which are shed as the plant grows older. Thus, in the specimen chosen for our illustration, these silky hairs have all fallen off, and nothing is seen of a hairy appearance except a coarser, somewhat glandular kind of hair, which remain, as shown by the plate, not only on the stem, but also on the peduncles or flower-stalks. These two conditions of our species must be borne in mind by the collector, since the hair is usually referred to as characteristic in the descriptions given by botanists.

Like most of the plants allied to the Asters, our species is an autumn bloomer. In Pennsylvania, for instance, it flowers early in September. There is such an abundance of yellow flowers of the asteraceous order at this season that there is hardly a desire for new species, especially as many of them have rather a weedy look. But the *Chrysopsis Mariana* has a very elegant habit of growth, and it ought, therefore, to be welcome in gardens, although we do not know of any attempts to cultivate it.

We know of no generally accepted English common name for the genus. Dr. Gray names it "Golden Aster," which is very pretty, but apt to be misunderstood, as these plants are not true Asters. "Gold Flower" would be quite appropriate, but unfortunately this has been given to a sort of poppy in California. We may get out of the difficulty, however, by translating "Aster," and so we shall call our flower the "Maryland Golden Star."

EXPLANATION OF THE PLATE. — 1. Stalk with flowers. — 2. Achene and pappus.

IRIS VIRGINICA.

BOSTON IRIS.

NATURAL ORDER, IRIDACEÆ.

IRIS VIRGINICA, Linnæus. — Stem round, slender, few-flowered; leaves linear, long; flowers beardless; ovary triangular, the side doubly grooved. Rhizoma fleshy. Stem smooth one to two lines in diameter, one foot to two feet high, branching at top and bearing two to six flowers. Bracts at the base of the branches withering. Leaves few, alternate, grass-like, six to ten inches long, amplexicaul. Sepals narrow, yellow, edged with purple. Petals linear-lanceolate. (Wood's *Class-Book of Botany*. See also Gray's *Manual of the Botany of the Northern United States*, and Chapman's *Flora of the Southern United States*.)

THE *Iris* is well known to all lovers of flowers. It occurs abundantly in a wild condition, and is a favorite in gardens; it has frequently been treated in poetry, painting, and sculpture, and plays an important part in history. In mythology it is said to have come from heaven. Iris was a messenger employed by Juno, and she is generally represented as sitting behind her mistress, her wings glittering like pearl, and radiant with all the colors of the rainbow. Her name, indeed, which literally means "eye of heaven," is the Greek word for rainbow.

The historical importance of the Iris is due to the fact that it became the national flower of France. As such it has acquired a world-wide reputation under the name of "Flower de luce" or "Fleur de lis," which is nothing but a corruption of "Fleur de Louis." But it had a political significance long before it was officially adopted by the kings of France. It was used as an emblem by the Byzantine emperors, although in what relation does not now appear, and the early Frankish kings of France also employed it. There is a legend, quoted by Prior, that a shield filled with these flowers was brought to King Clovis while engaged in battle, and King Louis VII adopted the flower, in June, 1137, as the national emblem of France, possibly to perpetuate the memory of some such event. The type of the French "Fleur de lis" is supposed to be the white Florentine Iris, which produces the orris-root of commerce.

There seems to be little doubt that the original "Flower of Louis" was an Iris. English writers, however, misled by the corrupted form of "Felur de lis," have imagined the flower to be a lily, and this idea is still current in the English literature of our own day. Even Webster's Dictionary has adopted this idea, for there we read: "Fleur-de-lis, French, flower of the lily, corrupted in English to flower-de-luce. The royal insignia of France, whether originally representing a lily or the head of a javelin, is disputed." Under "Flower-de-luce," however, where no allusion is made to the royal insignia of France, the same dictionary says that the word is identical with Iris, and quotes Spenser as an authority. But this quotation can hardly be called apt, if, as the dictionary intimates, the three terms, "Flower-de-luce," "Flower of the Lily," and "Iris," are to be looked upon as identical. Spenser, if we may judge from the following lines, was evidently quite well aware of the difference between the "Fleur-de-lis" and the lily:—

> "Strow me the grounde with Daffadown-Dillies,
> And Cowslips and Kingcups, and loved Lillies;
>   The pretty Paunce,
>   And the Chenisaunce
> Shall match with the fayre Floure Delice."

On the other hand it must be admitted that Shakespeare, who frequently refers to the Flower de luce, evidently regards it as a true lily. Thus he makes Perdita say in the "Winter's Tale":

> "Bold oxlips and
> The crown imperial; lilies of all kinds,
> The flower de luce being one! Oh, these I lack
> To make you garlands of; and, my sweet friend,
> To strew him o'er and o'er."

Some commentators think that Shakespeare merely classed the Iris with the lilies, but a contemporary of the poet refers to the "Flower de luce" in a manner which makes it unmistakable that the white lily was meant, describing it as having "six leaves whiter than snow, and in the middle the pretty little golden hammers."

Like so many others of the earliest known of our native flowers, our present species came to the botanists of Europe from Virginia, and was therefore named *Iris Virginica* by Linnæus. Pursh also found the plant during his wanderings, and, supposing it to be different from that described by Linnæus, named it *Iris prismatica*, in allusion to the prismatic shape of the ovary (Fig. 4). The Linnæan name, however, prevails, as *I. Virginica* and *I. prismatica* are now believed to be identical, although Mr. Baker, author of a monograph on Iridaceæ, and the most recent authority on these plants, maintains that the *I. Virginica* of Linnæus has nothing to do with our plant, being, in his opinion, only a variety of *I. versicolor*, and that, therefore, our Boston Iris should be called by Pursh's name, *I. prismatica*. But however this may be, it certainly cannot be denied that the generally accepted botanical name of our plant gives no idea of its geographical range, as the species is northern rather than southern. Dr. Chapman embraces it in his "Flora of the Southern United States," and locates it in "swamps, North Carolina, Tennessee, and northward." Prof. Wood says it is found from "Massachusetts to New Jersey." It is also found in Maine, and extends west to Lake Michigan. It might be looked for in the northern parts of Ohio and Indiana, but it is not in any collector's lists from these states that we know of. The popular name of the plant is the "Boston Iris," and this is much more appropriate, in reference to its geographical centre, than "Virginian Iris," which name it also sometimes receives.

The place of growth of the Boston Iris is generally in swamps. In New Jersey and Delaware it is often found blooming in very dry places, but the nature of these places makes it evident that water stands in them in winter. All the authors who mention it, speak of it as growing in wet or muddy places, with the exception of Mr. Ruger, who, in a note to the "Bulletin of the Torrey Botanical Club," in the volume for 1875, says that it grows on rocks at New Durham, in the State of New York, in company with *Silene inflata*. But whatever may be the circumstances under which it is forced to exist in a state of nature, there is no doubt that it prefers dry, rich garden ground to the swampy places in which it is originally found. Our Boston Iris is, indeed, one of the prettiest of cultivated plants. It blooms in June, and the flowers follow one another in close succession, keeping up the display for several weeks. The flowers produce seed in great abundance, and seedlings could no doubt be easily raised, but the plants can be propagated more readily by dividing the rhizomas or creeping stems. In English gardens our species was under cultivation before the year 1758.

The fertilization of the plant is a very interesting process to the student. From the arrangement of the stamens and pistils, it might be supposed that its pollen cannot reach the stigma without external aid. But the writer of this, for the purpose of keeping off the insects, placed fine gauze bags over some flowers which were about to expand, and yet these flowers produced perfect seed as well as those which had not been protected. We can infer from this that there is something still to be learned in regard to the fertilization of our species.

The flower stem has a much more branching character than the size of our page would permit us to show, but the peculiarly wavy or twisted growth of the branchlets, which, together with the delicate, narrow leaves, is very characteristic of this species, is well shown on the plate.

Our drawing is from a Massachusetts specimen, kindly furnished by Mr. Jackson Dawson.

EXPLANATION OF THE PLATE. — 1. Rhizoma, with a primary and secondary terminal growth, from the latter of which the flower-stem will grow the next year. — 2. Branchlet, showing flower in bloom with an unopened bud. 3. — Branchlet, showing that the first flower is faded before the second is ready to expand. — 4. Cross section of the ovary.

www.ingramcontent.com/pod-product-compliance
Lightning Source LLC
Chambersburg PA
CBHW032056220426
43664CB00008B/1025